The Hype about Hydrogen

About Island Press

Island Press is the only nonprofit organization in the United States whose principal purpose is the publication of books on environmental issues and natural resource management. We provide solutions-oriented information to professionals, public officials, business and community leaders, and concerned citizens who are shaping responses to environmental problems.

In 2004, Island Press celebrates its twentieth anniversary as the leading provider of timely and practical books that take a multidisciplinary approach to critical environmental concerns. Our growing list of titles reflects our commitment to bringing the best of an expanding body of literature to the environmental community throughout North America and the world.

Support for Island Press is provided by the Agua Fund, Brainerd Foundation, Geraldine R. Dodge Foundation, Doris Duke Charitable Foundation, Educational Foundation of America, The Ford Foundation, The George Gund Foundation, The William and Flora Hewlett Foundation, Henry Luce Foundation, The John D. and Catherine T. MacArthur Foundation, The Andrew W. Mellon Foundation, The Curtis and Edith Munson Foundation, National Environmental Trust, National Fish and Wildlife Foundation, The New-Land Foundation, Oak Foundation, The Overbrook Foundation, The David and Lucile Packard Foundation, The Pew Charitable Trusts, The Rockefeller Foundation, The Winslow Foundation, and other generous donors.

The opinions expressed in this book are those of the author(s) and do not necessarily reflect the views of these foundations.

To Patricia

The Hype about Hydrogen

ISLAND PRESS WASHINGTON COVELO LONDON

H

Fact and Fiction in the Race to Save the Climate

Joseph J. Romm

Library of Congress Cataloging-in-Publication Data

Romm, Joseph J.

The hype about hydrogen : fact and fiction in the race to save the climate / Joseph J. Romm.

    p. cm.

Includes bibliographical references and index.

1. Hydrogen—Research. 2. Fuel cells—Research. 3. Hydrogen as fuel—Economic aspects. 4. Global warming—Prevention. I. Title.

TP261.H9.R65 2004

333.79'68—dc22

2003021418

Printed on recycled, acid-free paper

Design by Teresa Bonner

Manufactured in the United States of America

10 9 8 7 6 5 4 3 2

Contents

Foreword

Just as day follows night, human attention is drawn to energy, probably the most important bounty of creation and enabler of the evolution of Earth's biosphere. Energy drives the tides, provides fresh water, conditions the air and climate, and nourishes the biosphere, which in turn nourishes humanity. We earthlings enjoy the fruits of diverse forms of energy in countless ways, especially now that advancing technology allows us to readily manipulate it.

The energy-based industrial and scientific revolution, which continues today, has given us massively magnified powers over the "forces of nature." Disease has been reduced; food and shelter have become more readily available; transportation and communication have leaped forward. Over time, this progress has accelerated the growth of population and the consumption of food, timber, minerals, and aquatic life. Billions of people are now able to live lives of material splendor. With a flick of the switch, we can illuminate where and when we wish, heat and cool our structures, move about (physically and figuratively) with unbelievable speed and freedom, dispose of our wastes, treat our sick, and explore the universe.

At the same time, this extraction and use of energy are at the heart of a host of problems, such as air pollution and its adverse effects on health and infrastructure, degradation of water quality, despoliation of the land from mining and waste disposal, and alteration of Earth's entire climate. The chances of continuing our economic progress hang on the achievement of unprecedented advances in science and technology, as well as a revolution in the way we measure progress.

Evidence of the challenge abounds. Despite concerted, sustained, and fruitful efforts, human population still expands at 70 million people per year, and debilitating sprawl is everywhere. Earth's climate is suddenly showing deeply disturbing changes for the worse. Petroleum production is peaking now, and the production peak for natural gas will follow within several decades, threatening the end of our long joyride with "cheap energy." A comic strip character once observed, "From here on down, it's going to be uphill all the way." I get that feeling when I think of the human bolide spinning its way into the twenty-first century.

How shall we make our future secure? Unlike the case in Disney's world, wishing will *not* make it so. And that brings us to the subject of this book. Sadly, the general public and most politicians lack sufficient knowledge to judge whether a technical proposal to set us on a better course to the future is plausible or whether it is hype—and this is certainly true when it comes to hydrogen. But imagination tempered with sound arithmetic and sustained support of research and development can result in nearly miraculous outcomes.

This is an important and opportune book. Joe Romm deftly applies his keenly developed technical knowledge, political experience, and analytical mind to a frank and sensible appraisal of the chances to transform our energy system over the next half-century or so into one that will employ hydrogen as a key energy carrier. Properly devised, this may allow major advances in the provision of energy services and improved protection of health and environment. The path *is* possible, but it will require a major, broad-based, and sustained research and development investment in both the public and private sectors.

Dr. Romm puts the already worn cliché "hydrogen economy" into a refreshing framework. For example, to point out one misconception he clarifies, hydrogen is not an energy *source* except in the futuristic realm of nuclear fusion, but it is a potent *carrier* of energy, albeit with the enormous disadvantage of having low energy density and not being readily storable.

Romm also reminds us that many decades will be required for the transition to an energy system with no net CO_2 emission. If we hope to be using hydrogen in a major way by 2050, we need to start now to devise ways to produce, convert, and carry it. Meanwhile, there are alternatives that can be employed more quickly, including major increases in efficiency of energy use in all sectors.

Given our current choices and policies, I am drawn to the observation that "mankind would rather commit suicide than learn arithmetic." We must gain enough wisdom to be able to be "emotionally moved by statistics." In the early days of complex research on controlled thermonuclear fusion, the effort was dubbed "Project Sherwood," alluding to the work of Robin Hood to rob the rich (deuterium in water) in order to pay the poor (low-cost energy). It also was interpreted as "Sherwood is (sure would be) nice if it would work!" It is a lot easier for the public to be persuaded by soothsayers that the fabled "hydrogen economy" is nearly here than for the science and technology community to make that possible!

The time is at hand to come to grips with the real challenge—and opportunity—before us of committing our creativeness and financial resources to the long and arduous task of safely navigating our way through the next hundred years. Romm's book transports the reader a hopeful (but nearly tearful) distance in coming to terms with energy realities and challenges for the twenty-first century.

John H. Gibbons
Science Advisor to the President, 1993–1998
Director, congressional Office of Technology
Assessment, 1979–1993

Introduction

Imagine a world in which you can drive your car to work each day without consuming any oil or producing any pollution. When you park your car at work or at home, you hook it up to the power grid, generating pollution-free electricity for your community. And, as part of the deal, you get money back from your utility.

You are living in the hydrogen economy, a high-tech Eden. Is it too good to be true? Will it happen in your lifetime?

The environmental paradise of a hydrogen economy rests on two pillars: a pollution-free source for the hydrogen itself and a device for converting it into useful energy without generating pollution. Let's start with the fuel cell—a small, modular electrochemical device, similar to a battery, but which can be continuously fueled. For most purposes, think of a fuel cell as a black box that takes in hydrogen and oxygen and puts out water plus electricity and heat, but no pollution whatsoever.

The first commercial stationary fuel cell was introduced in the early 1990s by United Technologies Corporation. Since fuel cells have no moving parts, they hold the promise of high reliability, and since power outages had caused countless business disruptions in the late 1990s, the product seemed like a sure winner.

These fuel cells were first used to provide guaranteed ultra-reliable power in the Technology Center of the First National Bank of Omaha. The center processes credit card orders from all over the country. "A single major retail client can lose as much as $6 million an hour if the Center's power fails and orders are not processed," reported Thomas Ditoro, the project's electrical engineer. In the bank's previous facility, its customers had experienced substantial losses from a power outage and failure of a battery backup system. That is why the First National Bank installed the most reliable electric power source it could find, a system developed by SurePower Corporation combining fuel cells with other advanced energy and electronic devices.[1]

The bank needed to maximize the availability of its computer system to protect existing clients while attracting new ones. A traditional system, combining an uninterruptible power supply (UPS) with power from the electric grid and backup diesel generators, would have more than a 63 percent probability of a major failure over its 20-year life. The SurePower system has less than a 1 percent chance of a major failure in 20 years. This may well be the difference between business success or bankruptcy.

After the system was installed in mid-1999, the bank used its new ultra-reliable power as a key feature in its marketing campaign, and as a result it has increased its market share. Dennis Hughes, the bank's lead property manager, said that the system's high reliability "isn't a luxury for us" but rather is a "competitive advantage. With SurePower, First National can raise our customers' service expectations while generating higher revenues." Although the initial cost of the fuel cell system was higher than that of the traditional UPS system, the life-cycle costs were much lower—a winning choice in every way.[2]

I started working with SurePower during this project and performed an environmental analysis of the system. I found many benefits other than high reliability and low life-cycle cost. Compared with a traditional system using a UPS and the electric grid, the SurePower system had a superior environmental performance. It had more than 40 percent lower emissions of carbon dioxide (CO_2, the primary greenhouse gas) and less than one one-thousandth the emissions of other air pollutants. Intrigued, I became an investor in the company, and later I helped it make an advance sale of the greenhouse gas credits that would be created by its next pollution-reducing project.

And yet *for more than four years after* its highly successful First National Bank of Omaha project, SurePower still was not able to sell a second high-reliability system to any other customer. Moreover, the manufacturer of the fuel cells themselves, UTC Fuel Cells, had only limited success selling the product for other applications and is phasing them out to pursue a different fuel cell technology. Exploring the complicated reasons for this unexpected business outcome (see Chapters 2 and 3) will help explain the challenges to, as well as the benefits of, accelerating the commercialization of stationary fuel cells.

Assuming we succeed with fuel cells, we must still find affordable, pollution-free sources for hydrogen to achieve a true hydrogen economy. We are a long way from finding them. The person credited with originating the phrase "hydrogen economy" in the early 1970s, Australian electrochemist John Bockris, wrote in 2002, "Boiled down to its minimalist description, the 'Hydrogen Economy' means that hydrogen would be used to transport energy from renewables (at nuclear or solar sources) over large distances; and to store it (for supply to cities) in large amounts."[3] Our cars, our homes, our industries would be powered not by pollution-generating fossil fuels—coal, gas, and oil, much of which is imported from geopolitically unstable regions—but by hydrogen from pollution-free domestic sources.

Unfortunately, the costs of producing hydrogen from renewable energy sources are extraordinarily high and likely to remain so for decades, given current U.S. energy policies. Virtually all hydrogen today is produced from fossil fuels in processes that generate significant quantities of greenhouse gases. Although a number of people have characterized the hydrogen economy as being just around the corner, what they are actually promoting is an economy built around hydrogen made from natural gas and other polluting fossil fuels.

Running *stationary* fuel cells on hydrogen produced from natural gas makes a great deal of sense and seems likely in the near future (see Chapters 2 and 3). But a transportation system based on hydrogen will be much slower in coming and more difficult to achieve than is widely appreciated. The technological challenges are immense. More important, *fueling cars with hydrogen made from natural gas makes no sense, either economically or environmentally* (see Chapters 6 and 8). The rapidly growing threat of global warming demands that hydrogen for cars be produced from sources that do not generate greenhouse gases.

Energy and Climate

Our energy choices are inextricably tied to the fate of our global climate. The burning of fossil fuels emits CO_2 into the atmosphere, where it builds up, blankets the planet, and traps heat, accelerating global warming (see Chapter 7). Earth's atmosphere now contains more CO_2 than at any time in the past 420,000 years—leading to rising global temperatures, more extreme weather events (including floods and droughts), sea level rise, the spread of tropical diseases, and the destruction of crucial habitats such as coral reefs.

The world's energy system is sometimes compared to an aircraft carrier to describe how hard it will be to change its direction. A more apt analogy is a coal-powered locomotive—and we need a new engine, new fuel, and even new tracks. The first problem is that carbon-emitting products and facilities have a very long lifetime: Cars last 13 to 15 years or more; coal plants can last 50 years. The second problem is that, once emitted, CO_2 lingers in the atmosphere, trapping heat for more than a century. These two facts together make it urgent that we avoid constructing another massive and long-lived generation of energy infrastructure that will cause us to miss the window of opportunity for carbon-free energy until the twenty-second century.

The International Energy Agency (IEA) projects that coal generation will double between 2000 and 2030. The projected new plants would commit the planet to total CO_2 emissions of some 500 billion metric tons over their lifetime. This "amounts to half the estimated total cumulative carbon emissions from all fossil fuel used globally over the past 250 years!" observed David Hawkins, director of the Natural Resources Defense Council's Climate Center, in testimony to the United States Congress in June 2003.[4]

The scientific community is increasingly speaking with one voice to warn us that, on such a path, even the best-case scenario for climate change is grim. A plausible worst-case scenario is an irreversible catastrophe. We are in the doubly dangerous situation of facing global climate change that may be much more extreme and may occur much more quickly than was expected just a few years ago (see Chapter 7). As the National Academy of Sciences explained in 2002, "the new paradigm of an abruptly changing climatic system has been well established by research over the last decade, but this new thinking is little-known and scarcely appreciated in the wider community of natural and social scientists and policy-makers."[5]

The path set by the current energy policy of the United States and the developing world will dramatically increase greenhouse gas emissions over the next few decades, which will force sharper and more painful reductions in the future when we finally do act. In the United States, the transportation sector alone is projected to generate nearly half of the 40 percent rise in CO_2 emissions forecast for 2025, which is long before hydrogen-powered cars could have a positive effect on greenhouse gas emissions (see Chapter 8).

Our current policy ignores clean energy technologies available today. Worse, our status quo energy policy works only if new technologies under development today become competitive in the marketplace quickly *and* if global warming is mild compared with most leading projections. Neither is likely to be the case. Thus, the path we are on is fraught with unnecessary risks.

A Realistic View

Many clean energy companies were overhyped in the late 1990s, in part because of a strange myth that, because the Internet and related information technology equipment supposedly consumed a great deal of electricity, the rapid growth in the Internet would lead to rapid growth in electricity demand. This, it was argued, would benefit all technologies that provide electricity but especially those that could provide reliable power, such as fuel cells. Several major brokerage firms released their own "analyses" repeating this myth and touting a variety of energy technology stocks. The myth was utterly refuted by the Lawrence Berkeley National Laboratory, the Rand Corporation, and my own Center for Energy and Climate Solutions, among others: The Internet is not, in fact, a big electricity draw, and its growth has little, if any, effect on overall growth in electricity demand (see Chapter 3).

The near-term prospects for fuel cell vehicles were also overhyped in the late 1990s. In November 2002, a major study titled "Hybrid & Competitive Automobile Powerplants" concluded, "The industry is currently experiencing a backlash to the 'just around the corner' hype that has surrounded the automotive fuel cell in recent years."[6]

Not surprisingly, from fuel cells to microturbines, many stocks that soared in the NASDAQ boom plummeted in the bust. For instance, the company that assumed a leadership role in transportation fuel cells in the 1980s and 1990s, the company with the most patents and the most major deals with automakers, is Ballard Power Systems Inc. This Canadian company was even the subject of a very favorable 1999 book, *Powering the Future,* in which executives were quoted as assuring profitability within a year or two. The stock price soared in the late 1990s. By December 2002, the price had dropped back to 1997 levels. The company announced that it was laying off 400 people, one-quarter of the workforce, and did not expect to achieve profitability for five years.[7] And yet Ballard remains one of the leaders in fuel cells. Commercializing new energy technologies is much harder than is widely realized—a key theme of this book.

For a number of years, I have divided my time between working with small companies trying to market the next breakthrough technology and helping Fortune 500 companies design strategies to cut energy costs and reduce greenhouse gas emissions. This work, together with my earlier time at the U.S. Department of Energy (see Chapter 1), has taught me two large lessons about the marketplace.

First, most companies are very conservative about purchasing and deploying new technology. A small number of firms are aggressive first adopters, but the vast majority buy only products that have both a good commercial track record and a very rapid payback. Even many brand-named companies will invest only in an energy technology that pays for itself in energy savings within about a year.

Second, those in the public or private sector who advocate new technologies tend to overestimate how rapidly they will achieve their performance and cost goals while underestimating what the competition will do. Renewable energy still suffers from the marketplace perception that it has failed to deliver on promises made in the 1970s, although for more than a decade now both solar power and wind power have been growing rapidly, and most renewable energy technologies have met or exceeded their cost and performance goals. Nonetheless, renewable energy does not now provide a bigger share of U.S. energy mainly because the competition got tougher. Fossil fuel technologies in particular now have both reduced costs and reduced pollution (see Chapter 8).

Hydrogen faces a similar set of obstacles. Yet numerous studies of a hydrogen economy rely on assumptions that are overly optimistic. For instance, many analyses assume that the total delivered cost for hydrogen need be reduced only to a level at which it is twice that of gasoline (for an equivalent amount of energy delivered). The argument is made that hydrogen fuel cells are twice as efficient as gasoline internal combustion engines, so the fuel can be twice as expensive and consumers will still end up paying the same total fuel bill. But that is comparing a future technology with a current technology—and not even the best current technology. Hybrid gasoline-electric vehicles *today*, such as the Toyota Prius, are already much more efficient than traditional internal combustion engine vehicles and nearly as efficient as *projected* fuel cell vehicles (assuming fuel cells achieve their performance targets). A hybrid diesel-electric vehicle would have about the same overall efficiency as a fuel cell vehicle. That's tough competition for hydrogen.

Given the numerous large roadblocks that hydrogen fuel cell vehicles must overcome to become competitive products, and given the history of other advanced energy technologies, we should avoid both overoptimistic assumptions about new technologies and underestimation of the competition. Indeed, considering the tens of billions of dollars for infrastructure that the government (and hence U.S. taxpayers) will have to devote to bring about a hydrogen economy, conservative assumptions are essential. The energy and environmental problems facing the nation and the world, especially global warming, are far too serious to risk making major policy mistakes that misallocate scarce resources.

In this book, my aim is to be realistic, using analysis that is neither overly optimistic nor overly pessimistic. In almost every case in which I cite a cost or make a simple calculation, there will be other ways to do the analysis based on different assumptions, projections of future technological breakthroughs, or estimates of how mass production of existing technology could dramatically cut costs. I am very hopeful that the sunnier predictions will ultimately prove true, but our limited experience with commercializing fuel cells provides a multi-decade lesson in high-tech humility. And our recent experience in trying to accelerate the introduction of alternative fuel vehicles provides a lesson in how difficult it will be to rapidly change gasoline-powered cars and the gasoline infrastructure (see Chapters 5 and 6). One hard lesson learned is that over-hyping new technologies ultimately ends up slowing their success in the market.

Unfortunately, the usual sources for good information have often been unreliable, a testimony to the enormous difficulty of analyzing the many factors involved in the transition to a pollution-free hydrogen economy. Most of the articles and books on the subject in recent years—including articles in such prestigious publications as *Technology Review*, *Wired* magazine, and the *Atlantic Monthly*—fail to distinguish between likely scenarios for the future and unlikely ones. They often contain serious errors and misleading statements.

Just as importantly, major corporations such as General Motors continue to overhype the near-term prospect for hydrogen cars. GM

is wildly overestimating the speed of successful mass-market introduction of hydrogen cars, which it says will start around 2010, while
underestimating the competition from hybrids. GM is spending a
large fraction of its research budget on hydrogen-powered cars—
which the analysis in this book suggests is a strategic mistake.[8]

My goal here is threefold: to lay out the key issues in the transition to a hydrogen economy, to sort out the plausible scenarios from
the less plausible ones, and to explain how current policies could
delay that transition 10 to 20 years or more. Since no single person
can be an expert in all relevant areas, I have sought out a spectrum
of views from dozens of scientists, engineers, government officials,
entrepreneurs, environmentalists, and policy analysts.

This book makes the case that hydrogen vehicles are unlikely to
achieve even a 5 percent market penetration by 2030. And this in
turn leads to the book's major conclusion for all readers, from pol-
icymakers to corporate executives to investors to anyone who cares
about the future of the planet: *Neither government policy nor business
investment should be based on the belief that hydrogen cars will have
meaningful commercial success in the near- or medium-term.*

CHAPTER 1

Why Hydrogen? Why Now?

When I first came to the U.S. Department of Energy (DOE) in
1993 to help oversee research and development (R&D) in
clean energy, hydrogen R&D did not even have its own separate
budget line but instead was nestled inside the renewable energy
budget. For the previous decade, hydrogen research funding had lan-
guished in the $1–$2 million per year range, some one one-hundredth
of 1 percent of the overall departmental budget—a penny in every $100.

Only ten years later, all the major car companies had hydrogen
vehicle programs, the major oil companies had hydrogen production programs, dozens of new companies had been formed to
develop hydrogen-related technologies with venture capital funding, and President George W. Bush had announced a major hydrogen initiative in his January 2003 State of the Union address:

> Tonight I'm proposing $1.2 billion in research funding so that
> America can lead the world in developing clean, hydrogen-
> powered automobiles. A single chemical reaction between hydro-
> gen and oxygen generates energy, which can be used to power a
> car—producing only water, not exhaust fumes. With a new
> national commitment, our scientists and engineers will overcome
> obstacles to taking these cars from laboratory to showroom, so

that the first car driven by a child born today could be powered by hydrogen, and pollution-free.[1]

What caused this sea change in ten short years? I believe there are three reasons: a series of technological advances in the fuel cells most suitable for cars; growing concern about a variety of energy and environmental problems, especially global warming; and advances in technologies needed for greenhouse-gas-free hydrogen production.

Advances in Transportation Fuel Cells

Fuel cells are one of the Holy Grails of energy technology (see Chapter 2). They are pollution-free electric "engines" that run on hydrogen. Unlike virtually all other engines, fuel cells do not rely on the burning of fossil fuels. Hence, they produce no combustion by-products, such as oxides of nitrogen, sulfur dioxide, or particulates—the air pollutants that cause smog and acid rain and that have been most clearly documented as harmful to human health.

Fuel cells have been reliably providing electricity to spacecraft since the 1960s, including the Gemini and Apollo missions as well as the space shuttle. The leading manufacturer of fuel cells for the National Aeronautics and Space Administration (NASA), United Technologies Corporation, has sold commercial units for stationary power since the early 1990s, with more than 200 units in service.

But finding a fuel cell with the right combination of features for powering a car or truck has proved much more difficult. Why is that hard? To begin with, you need a fuel cell that is lightweight and compact enough to fit under the hood of a car but that can still deliver the power and acceleration drivers have come to expect. You also need a fuel cell that can reach full power in a matter of seconds after start-up, which rules out a variety of fuel cells that operate at very high temperatures and thus take a long time to warm up. You also need cost and reliability comparable to that of the gasoline-powered internal combustion engine, which is an exceedingly mature technology, the product of more than a hundred years of development and real-world testing in hundreds of millions of vehicles.

Further, these hardware hurdles are all quite separate from the way in which the fuel for fuel cells—hydrogen—would be produced and delivered to the vehicle. Hydrogen, first and foremost, is not a primary fuel, like natural gas or coal or wood, which we can drill or dig for or chop down and then use at once. Hydrogen is the most abundant element in the universe, true enough. But on Earth, it is bound up tightly in molecules of water, coal, natural gas, and so on. To unbind it, a great deal of energy must be used.[2]

For all these reasons—plus the sharp drop in both the price of oil and government funding for alternative energy—hydrogen fuel cell vehicles received little attention through most of the 1980s. Still, a few government and industry laboratories (together with a small and ardent group of hydrogen advocates and state energy experts) kept plugging away, particularly on proton exchange membrane (PEM) fuel cells.

PEM fuel cells were developed in the early 1960s by the General Electric Company for the Gemini space program. Fuel cells require catalysts to speed up the electrochemical reaction, and PEM fuel cells use platinum, a very expensive metal. An early 1981 analysis for the DOE had presciently argued that PEM fuel cells would be ideal for transportation if the catalyst loading could be significantly reduced.[3] By the early 1990s, Los Alamos National Laboratory (and others) did succeed in cutting the amount of platinum by almost a factor of ten, a remarkable improvement. This still did not make PEM fuel cells cost-competitive with gasoline engines—we are a long way away from that—but it did dramatically reinvigorate interest in hydrogen-powered vehicles because PEMs were exactly the kind of low-temperature fuel cell that could be used in a car.

In 1993, DOE funding for PEM fuel cells was just less than $10 million. Within days of my arrival, I was briefed on Los Alamos' PEM work and began pushing for increases in funding for PEM fuel cells as well as for the development of a transportation fuel cell strategy. President Bill Clinton's entire team was very supportive of R&D for fuel-efficient technologies, including PEMs. Funding for hybrid vehicles, including fuel cells, was significantly increased. So, too, was funding for hydrogen R&D.

In mid-1995, I moved to the DOE's Office of Energy Efficiency and Renewable Energy. I was principal deputy assistant secretary, the number two slot, in charge of all budget and technology analysis. In that capacity, I was able to work with other fuel cell advocates in and out of the administration, especially my DOE colleagues Brian Castelli and Christine Ervin, to keep the PEM fuel cell budget creeping upward even as the entire budget for the office was cut 20 percent by a 1995 Congress that was extremely skeptical of all energy R&D.[4]

It seemed likely that long before PEM fuel cells would be cost-effective for powering cars, they would be cost-effective for providing electricity and hot water to buildings. Yet Congress had repeatedly rejected our office's request to start a small ($1 million) program to advance the effort to put fuel cells into buildings. In 1997, when I was acting assistant secretary, we managed to launch the program for stationary PEM fuel cell research, the budget for which ultimately grew to several million dollars.

By 1998, the year I left the DOE, the hydrogen budget was ten times larger than in 1993, and the proposed PEM fuel cell budget was more than three times larger. The investment paid off: The cost of fuel cells (PEM and others) had steadily declined as performance increased. Harry Pearce, vice chairman of the General Motors Corporation, said at the North American International Auto Show in January 2000, "It was the Department of Energy that took fuel cells from the aerospace industry to the automotive industry, and they should receive a lot of credit for bringing it to us."[5]

There were promising developments in hydrogen production and storage. Hydrogen budgets were ballooning everywhere in a race for patents and products. William Clay Ford Jr., chairman of the Ford Motor Company, said in October 2000, "I believe fuel cells will finally end the 100-year reign of the internal combustion engine"—a poignant statement from the great-grandson of the man whose manufacturing innovations had begun that reign a century ago with the Model T.[6]

The federal government's increasing commitment to hydrogen and fuel cells, together with the technological successes already spawned by that funding, has spurred private sector interest, much as similar support does in medicine and national defense. Since the late 1990s, hundreds of millions of dollars from venture capitalists and investors in the stock market have flowed into start-up companies and divisions of existing companies, all working to develop hydrogen-related technologies and fuel cells, although, as discussed in Chapter 3, this investment funding has proved as erratic as the stock market.

The U.S. government was hardly the only source of funding for hydrogen and PEM fuel cells. Governments in Europe and Asia have major programs, as do Japanese car companies such as Toyota and Honda. Canada has a significant program because of the leadership of Ballard Power Systems Inc. in PEM technology.

Growing Energy Risks

Many other trends have driven the renewed interest in hydrogen. At the top of the list are worries about oil consumption and air pollution, including global warming. America's dependence on imported oil has accelerated since the mid-1990s, as many people predicted—including Charles Curtis, then deputy secretary of the DOE, and me in a 1996 *Atlantic Monthly* piece titled "Mideast Oil Forever?"[7] By 2002, we were importing more than half our oil, an outflow of $100 billion per year to foreign governments, including those in the politically unstable Persian Gulf region.

The terrible September 11, 2001, terrorist attacks heightened this concern. Less than two weeks later, the DOE was commenting publicly. "It is clear that our reliance on imported oil—56% of the oil we use—has complicated our response to the terrorist attack," noted David Garman, the Bush administration's assistant secretary for energy efficiency and renewable energy, on September 24, 2001. "There is also little doubt that some of the dollars we have exported in exchange for foreign oil have found their way into the hands of terrorists and would-be terrorists."[8]

These seismic problems, together with worldwide population growth, economic growth, and urbanization, will dramatically increase global oil consumption in the coming decades, especially in the developing world. If by 2050 the per capita energy consumption of China and India were to approach that of South Korea, and if the Chinese and Indian populations increase at currently projected rates, those two supergiant countries *by themselves* would consume more oil than the entire world did in 2003.[9]

Since oil is a finite, nonrenewable resource, analysts have attempted to predict when production will peak and start declining. Some believe this will occur by 2010. The Royal Dutch/Shell Group, probably the most successful predictor in the global oil business, adds fifteen to thirty years to that gloomy forecast. Worry about oil supplies is one of the factors behind Shell's growing research into hydrogen (see Chapter 7). This debate will not be resolved here, but it does appear credible that oil production will peak in the first half of this century and will possibly decline at a relatively rapid rate thereafter, even as demand increases. Thus, delaying action until we are past the peak may put us at significant risk.

Growing Environmental Risks

A whopping two-thirds of U.S. oil consumption is in the transportation sector, the only sector of the U.S. economy wholly reliant on oil. The energy price shocks of the 1970s helped spur growth in use of natural gas for home heating and drove the electric utility sector and the industrial sector to reduce their dependence on petroleum. But roughly 97 percent of all energy consumed by our cars, sport utility vehicles, vans, trucks, and airplanes is still petroleum-based.

Not surprisingly, a high priority of R&D funding by the United States—and by any country, state, or company that takes the long view—is to develop both more fuel-efficient vehicles and alternative fuels. Only a limited number of fuels are plausible alternatives for gasoline, and one enormous benefit of hydrogen over others is that it can be generated by a variety of different sources, thus potentially minimizing dependence on any one. Most important, hydrogen can be generated from renewable sources of energy such as wind power, raising the ultimate prospect of an inexhaustible, clean, domestic source of transportation fuel. Also, since fuel cells are more efficient than gasoline internal combustion engines, hydrogen fuel cell vehicles are, potentially, a double winner in the race to replace oil.

Hydrogen fuel cell vehicles would seem to be the perfect answer to our burgeoning and alarming dependence on imported oil. For some, like Peter Schwartz, chair of the Global Business Network, they are almost the deus ex machina—the quick, pure technological fix—that will avoid the need for difficult policy choices, such as federal mandates for increased vehicle efficiency.[10] That is overoptimistic hype, as we will see.

The pollution generated by internal combustion engine automobiles is another key reason why so many people are drawn to hydrogen fuel cell vehicles. The transportation sector remains one of the largest sources of urban air pollution, especially the oxides of nitrogen that are a precursor to ozone smog and the particulates that do so much damage to our hearts and lungs. Vehicle emissions of such pollutants, however, have been declining steadily, and, by 2010, federal and state standards will have made new U.S. cars exceedingly clean.

Yet, even as new internal combustion engine vehicles dramatically cut the emissions of noxious urban air pollutants by automobiles, their contribution to global warming has begun to rise. In the 1990s, the transportation sector saw the fastest growth in carbon dioxide (CO_2) emissions of any major sector of the U.S. economy. And *the transportation sector is projected to generate nearly half of the 40 percent rise in U.S. CO_2 emissions forecast for 2025.*[11]

When the United States takes serious action on global warming, the transportation sector will need to be a top priority. The two most straightforward ways to reduce vehicle CO_2 emissions are, first, by increasing the fuel efficiency of the vehicles themselves and, second, by using a fuel that has lower net emissions than gasoline. Again, the attractiveness of hydrogen fuel cell vehicles is that they afford the possibility of pursuing both strategies at the same time: Fuel cells are more efficient than traditional internal combustion engines, and hydrogen, when produced from renewable energy sources, would create no net greenhouse gas emissions.

Hydrogen without Greenhouse Gases

The possibility that hydrogen and fuel cells could play a key role in combating pollution, particularly global warming, is, I believe, the strongest argument for expanded efforts in research and development. John Heywood, director of the Sloan Automotive Laboratory at the Massachusetts Institute of Technology, argues, "If the hydrogen does not come from renewable sources, then it is simply not worth doing, environmentally or economically."[12]

The idea that hydrogen could be generated without releasing any pollution is not a new one. In 1923, John Haldane, who later became one of the century's most famous geneticists, gave a lecture predicting that Britain would ultimately derive its energy from "rows of metallic windmills" generating electricity for the country and, when there was excess wind, producing hydrogen. "Among its more obvious advantages will be the fact that . . . no smoke or ash will be produced."[13]

The problem with the vision of a pure hydrogen economy has been that, until recently, most greenhouse-gas-free sources for hydrogen have been far too expensive to be practical. Haldane himself was imagining a future "four hundred years hence." Even today, nuclear, wind, and solar electric power would produce hydrogen that is far more expensive than hydrogen from fossil fuels. But for more than two decades, renewable energy, especially wind and solar energy, has been declining in price sharply. That has created a renewed interest in renewable hydrogen, although it will still be two or more decades before this is a competitive way to generate hydrogen.

There is another, more unexpected possible source of greenhouse-gas-free hydrogen: fossil fuels. In the mid-1990s, Princeton University professor Bob Williams (and others) produced detailed reports arguing that fossil fuels could be both a cost-effective and an environmentally benign source of hydrogen *if* the CO_2 released during the production process could be captured and stored in underground geologic formations so that it would not be released into the atmosphere and thereby accelerate global warming. His briefings to DOE officials and others in government were a major reason why the department launched a major effort to explore this possibility. Today, carbon capture and storage is the subject of considerable research as well as demonstration projects around the globe and is widely seen as a potentially critical strategy for addressing global warming in the longer term (see Chapters 7 and 8).

With ongoing advances in transportation fuel cells and pollution-free hydrogen production, hydrogen vehicles would seem to be the perfect answer to global warming. Yet one of the conclusions of this book—the one that surprised me the most, as it was not my view before starting the current research—is that *hydrogen vehicles are unlikely to make a significant dent in U.S. greenhouse gas emissions in the first half of this century*, especially if U.S. energy policy is not significantly changed (see Chapter 8). Still, hydrogen-fueled *stationary* power plants could be critical in reducing greenhouse gas emissions much sooner. Further, hydrogen may well be the essential vehicle fuel in the second half of this century if we are to achieve the very deep reductions in CO_2 emissions that will almost certainly be needed then or if we are past the peak of oil production.

We are not used to thinking or planning in such giant, multi-decade time steps. But then again, we have never faced such a giant problem as global warming.

The Long Transition to a Hydrogen Economy

The term "hydrogen economy" describes a time when a substantial fraction of our energy is delivered by hydrogen made from sources of energy that have no net emissions of greenhouse gases.[14] These would include renewable sources of energy, such as wind power and biomass (e.g., plant matter), but it could also include the scenario of converting fossil fuels into hydrogen and CO_2 and then permanently storing the carbon. It could also include generating hydrogen from nuclear power, should that prove practical.

We are unlikely to know whether a hydrogen economy is practical and economically feasible for at least one decade and possibly two or even more. A hydrogen economy would require dramatic changes in our transportation system because, at room temperature and pressure, hydrogen takes up three thousand times more space than gasoline containing an equivalent amount of energy. We will need tens of thousands of hydrogen fueling stations, and, unless hydrogen is generated on-site at those stations, we will also need a massive infrastructure for delivering that hydrogen from wherever it is generated.

Substantial technological and cost breakthroughs will be needed in many areas, not the least of which is fuel cells for vehicles. In 2003, fuel cell vehicles cost $1 million each or more. Were we to build a hydrogen infrastructure for fueling vehicles, the total delivered cost of hydrogen generated from fossil fuel sources would likely be at least triple the cost of gasoline for the foreseeable future.[15] Hydrogen generated from renewable energy sources would be considerably more expensive. Hydrogen storage is currently expensive, inefficient, and subject to onerous codes and standards. The DOE does not foresee making a decision about commercializing fuel cell vehicles until 2015.[16] One detailed 2003 analysis of a hydrogen economy by two leading European fuel cell experts concluded, "The 'pure-hydrogen-only-solution' may never become reality."[17]

And if the imposing technical and cost problems can be substantially solved, we will still have an imposing chicken-and-egg problem: Who will spend hundreds of billions of dollars on a wholly new nationwide infrastructure to provide ready access to hydrogen for consumers with fuel cell vehicles until millions of hydrogen vehicles are on the road? Yet who will manufacture and market such vehicles—and who will buy them—until the infrastructure is in place to fuel those vehicles? A 2002 analysis by Argonne National Laboratory found that "with current technologies, the hydrogen delivery infrastructure to serve 40% of the light duty fleet is likely to cost over $500 billion."[18] I fervently hope to see an economically, environmentally, and politically plausible scenario for bridging this classic catch-22 chasm; it does not yet exist.

So, despite much hype to the contrary, a hydrogen economy is a long way off. Widespread use of fuel cells, particularly for stationary applications, may be just around the corner, however, so they are the subject of the next two chapters.

CHAPTER 2

Fuel Cell Basics

Fuel cells convert the energy of a chemical reaction between a fuel such as hydrogen and an oxidant such as oxygen directly into electric energy and heat. Since water (H_2O) is composed of hydrogen and oxygen, it is perhaps not surprising that fuel cells were invented by someone experimenting with decomposing water into oxygen and hydrogen using electricity.

In late-1830s London, Sir William Grove (1811–1896) figured out how to reverse the electrolysis process to generate electricity by combining oxygen and hydrogen.[1] Grove then quickly built the first fuel cell, which he called a "gaseous voltaic battery," or gas battery. He used electrodes made of platinum, which was known to catalyze the reaction between hydrogen and oxygen—that is, to help speed the reaction without itself changing in the process.

By 2003, more than 160 years after the first fuel cell was built, and after more than $15 billion in public and private spending, only one fuel cell with significant commercial sales existed, and purchasers of that product received a government subsidy to buy down the cost.[2] Moreover, the manufacturer of that specific product, UTC Fuel Cells, is phasing it out in favor of newer technologies. This chapter examines the fuel cells that are the subject of the most government and corporate interest today. It is for the more technically inclined, but to follow the reasoning in later chapters, you will not need to understand how fuel cells work. For most purposes, you can think of a fuel cell as a "black box" that takes in hydrogen and oxygen and puts out only water plus electricity and heat, though, as this chapter points out, not all these black boxes are the same, and some may be better suited to some purposes than others.

How a Fuel Cell Works

Fuel cells produce electricity, heat, and water by catalyzing the reaction of hydrogen and oxygen. Today, hydrogen is most easily and cheaply generated from natural gas in a process known as reformation (see Chapter 3). The reforming process can be performed *internally* for fuel cells that operate at very high temperatures—that is, high-temperature fuel cells can run directly on natural gas. But the low-temperature fuel cells ideal for transportation require an *external* reformer that delivers hydrogen of very high purity.

The fuel cell is composed of a positively charged electrode (the cathode) and a negatively charged electrode (the anode) separated by an electrolyte. Electrolytes can be made of many materials, including ceramics and plastic membranes. See Figure 2.1 for a schematic diagram of a PEM (proton exchange membrane or polymer electrolyte membrane) fuel cell producing electric power.[3]

The thin polymer (plastic) electrolyte prevents the flow of electrons between the anode and cathode. The electrolyte also prevents the flow of hydrogen, which is composed of one electron and one proton. Hydrogen ions, on the other hand, consist of only one proton and can pass through. In PEM fuel cells, the hydrogen passes over the anode, where it is split into electrons and hydrogen ions (protons) with the help of a platinum catalyst. The electrons pass through an external circuit, where their energy is used to run an electric device such as a motor, before migrating to the cathode. In this regard, a fuel cell is similar to a battery, one that can be continuously fueled. Once the positively charged hydrogen ions have passed through the polymer electrolyte at the center of the fuel cell, they are combined with the oxygen at the cathode to produce

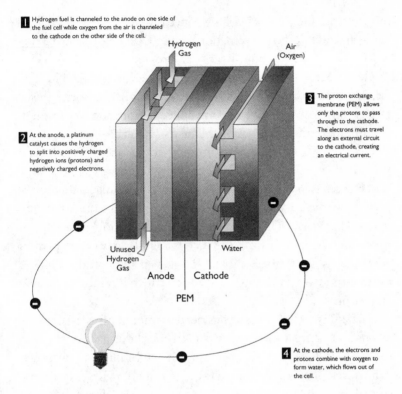

1 Hydrogen fuel is channeled to the anode on one side of the fuel cell while oxygen from the air is channeled to the cathode on the other side of the cell.

Hydrogen Gas

Air (Oxygen)

3 The proton exchange membrane (PEM) allows only the protons to pass through to the cathode. The electrons must travel along an external circuit to the cathode, creating an electrical current.

2 At the anode, a platinum catalyst causes the hydrogen to split into positively charged hydrogen ions (protons) and negatively charged electrons.

Unused Hydrogen Gas

Water

Anode

Cathode

PEM

4 At the cathode, the electrons and protons combine with oxygen to form water, which flows out of the cell.

FIGURE 2.1. Structure of a PEM fuel cell.

water, again facilitated by the platinum catalyst. Since each fuel cell produces less than one volt, fuel cells are connected in stacks to produce higher voltages.

The Different Types of Fuel Cells

The material used for the electrolyte determines many of the fuel cell's properties, including its operating temperature, and thus is used to distinguish the different types of fuel cells. All the leading fuel cells are discussed in the sections that follow.

Phosphoric acid fuel cells, as its name says, use phosphoric acid as the electrolyte—Grove's first fuel cell used sulfuric acid. They are the most mature fuel cell technology. Since the early 1990s, UTC Fuel Cells, formerly called International Fuel Cells, has been making what it calls the PC25 unit, which can provide 200 kilowatts (kW) of power. More than two hundred units have been placed in service, in places as diverse as Pittsburgh International Airport, a postal facility in Alaska, a New York City police station, and the Technology Center of the First National Bank of Omaha.[4]

PC25s operate at temperatures of about 300°F–400°F and require an external reformer. The electric efficiency of a phosphoric acid fuel cell is in the range of 30 to 35 percent. That is, only 30 to 35 percent of the total energy in the natural gas is converted into usable electricity.[5] The average U.S. central-station power plant producing electricity from fossil fuels is a similarly paltry 30 percent efficient. By contrast, new natural gas plants producing 100 megawatts (MW) are doing much better, with efficiencies exceeding 55 percent.

Each PC25 also supplies a considerable amount of heat—as much as 900,000 British thermal units (Btu) per hour at 140°F, which can be used for a variety of commercial and industrial purposes. Annualized, that is approximately the amount of space heating consumed by 100 households in the northeastern United States.[6] PC25s can also be configured to provide half of their heat at 250°F. In general, the temperature of the usable heat from a fuel cell is not the same as the temperature at which the fuel cell stack operates.[7] The heat that is rejected by the fuel cell stack is used by the overall fuel cell system for preheating air, raising steam, and preheating or vaporizing fuel.

The simultaneous generation of both electricity and heat is called *cogeneration,* or combined heat and power (CHP). A cogenerating PC25 can achieve an overall efficiency of 80 percent. Nevertheless, phosphoric acid fuel cells have had limited commercial success because they are much more costly than other potential small-scale power generation technologies. Other existing power generation technologies can deliver comparable amounts of electricity and heat with almost the same efficiency but for less than $1,000 per kW (see Chapter 3). Phosphoric acid fuel cells begin with a high cost,

$4,500/kW—and the total installed costs can be even higher, depending on the complexity of the project, the difficulty of integrating the cogenerated heat into the building or factory, the costs of connecting with the electric grid, and so on.

Phosphoric acid fuel cells rely on expensive components and, like PEMs, use platinum catalysts to accelerate the chemical reactions at the electrodes. Finally, they have not achieved the level of sales needed to significantly reduce manufacturing costs. For these reasons, UTC Fuel Cells is phasing out production of phosphoric acid fuel cells in favor of PEM fuel cell technology, which is likely to be significantly less expensive.

Molten carbonate fuel cells use liquid carbonate as the electrolyte. They have been the subject of research and development (R&D) for more than four decades and are currently in the early stages of commercialization by FuelCell Energy Inc. of Danbury, Connecticut, after an expenditure of some $200 million in R&D funding from the U.S. Department of Energy (DOE). These fuel cells have several distinct advantages over phosphoric acid fuel cells. First, because they operate at high temperatures (1,200°F or higher), they can use that heat internally to produce hydrogen directly from a variety of fuels, including natural gas, ethanol, and methanol, which is why the company calls its product Direct FuelCells. This makes them more efficient than lower-temperature fuel cells, such as phosphoric acid fuel cells and PEMs, which require an external reformer to generate hydrogen from fuels such as natural gas, a process that can be both costly and inefficient.

Second, molten carbonate fuel cells have electric efficiencies of 47 to 50 percent or more, which significantly reduces their fuel costs for stationary applications compared with both phosphoric acid and PEM fuel cells, whose overall efficiency when running on natural gas might not exceed 35 to 40 percent. Third, high temperatures allow relatively inexpensive nickel to be used as a catalyst rather than pricey platinum, which is required by the lower-temperature fuel cells. Fourth, these fuel cells are far more tolerant of carbon monoxide, which can "poison" the electrochemical reaction of PEM fuel cells. Finally, the high-temperature steam produced by molten carbonates is better suited for many commercial and industrial processes. As of 2003, the 250 kW unit marketed by FuelCell Energy is cogenerating heat with an exhaust temperature of about 650°F.

Molten carbonates, however, do have disadvantages. The fuel cell takes a considerable amount of time to reach its high operating temperature, so it is unsuitable for powering a car or truck. The liquid carbonate electrolyte is very corrosive, so there is some question about the lifetime these fuel cells will be able to achieve. They are also very bulky: The 250 kW units are the size of a railroad car and weigh about forty tons.

Molten carbonate fuel cells are currently being demonstrated at several sites around the world, including hotels in New York and New Jersey (see Chapter 3). In demonstration projects for 2003, molten carbonate fuel cells are expected to cost $4,500/kW—and, as with phosphoric acid fuel cells, total installed costs are typically higher, depending on the complexity of the project. The cost of molten carbonate fuel cells is expected to decrease significantly as production volume increases. By 2002, FuelCell Energy had the capacity to manufacture 50 MW of fuel cell production, yet, as of 2003, the plant is operating far below capacity because of a lack of customers. The company projects that when it can sell its current capacity, costs will be cut almost in half. Ultimately, the company's goal is 400 MW of production, which it believes will bring costs down to $1,200–$1,500/kW, the point at which business analysis suggests a significant entry market exists for an unsubsidized fuel cell. However, achieving that low price for fully installed systems may take many years.

The company is also pursuing a hybrid system in which the fuel cell would work together with a gas turbine. In the near term, the combined system could convert a remarkable 65 to 70 percent of the energy in natural gas into electricity in megawatt-sized power plants, and, ultimately, efficiencies of 75 percent may be achievable for larger systems in the 10 MW range. This is significantly more efficient than the most efficient electric power plants—large combined cycle gas turbines—which is particularly important, given concerns about natural gas prices (see Chapter 8).

Solid oxide fuel cells (SOFCs) use ceramic as the electrolyte. They operate at a very high temperature, up to 1,800°F, and can operate with air and natural gas (or other fuels) as direct inputs. In contrast to PEMs (in which positively charged hydrogen ions travel through the polymer membrane), SOFCs use negatively charged oxygen ions that travel through a porous cathode, then through the electrolyte, and finally through a porous anode, where they combine with the hydrogen to form water. The solid ceramic electrolyte is a hermetic barrier between the chemical reactants, so no hydrogen or water can reach the air side of the fuel cell, which simplifies operation.

Unlike molten carbonates, solid oxides use a hard ceramic electrolyte instead of a liquid. That means the fuel cell can be cast into a variety of useful shapes, such as tubes. With higher temperatures, SOFCs may be able to cogenerate steam at temperatures as high as 1,000°F. The Siemens Westinghouse Power Corporation has built the first advanced hybrid system, which combines a gas turbine with a tubular SOFC. As of 2003, the 220 kW hybrid system has operated in California for more than 2,000 hours with a respectable 53 percent efficiency, comparable to current combined cycle gas turbines. The ultimate goal is an efficiency of 70 percent or more.

Through most of the past two decades, SOFCs were thought to be best suited for very large, multi-megawatt central-station power generation. That thinking has changed with improvements in ceramic technology and an increased focus on distributed generation—smaller power plants built on the site of commercial buildings or industrial facilities (see Chapter 3). In 2001, the DOE launched a ten-year, $300 million effort to fund industry teams to develop smaller-scale SOFCs, called the Solid State Energy Conversion Alliance. The goal is to achieve an SOFC that costs $400/kW, one-tenth of current costs for phosphoric acid and molten carbonates.

This SOFC alliance is second only to the manufacturers of PEMs in the magnitude of effort put out to achieve commercialization at a price that would guarantee marketplace success. As of April 2003, six industry teams are being supported by the DOE with R&D funds to develop SOFCs. And there are many other companies here and abroad pursuing SOFCs, whereas only one company is currently commercializing molten carbonate fuel cells (and that company, FuelCell Energy, is also one of the six DOE-funded teams pursuing solid oxides). So even though molten carbonates have a several-year advantage in the commercialization process, solid oxides may well be the high-temperature fuel cell winner, especially given SOFCs' potential advantages over molten carbonates, such as manufacturers' ability to mass-produce ceramic and other components using techniques developed in the semiconductor industry.

Nevertheless, operating at very high temperatures brings problems that have yet to be fully solved: As the 2002 *Fuel Cell Handbook* notes, there are "thermal expansion mismatches among materials, and sealing between cells is difficult in the flat plate configurations."[8] The very fact that SOFCs are "the fuel cell with the longest continuous development period, starting in the late 1950s" also underscores the difficulty SOFCs have had over the long decades in solving the technical and cost obstacles that block the way to delivering a commercial product.

Proton exchange membrane (PEM) fuel cells, also known as polymer electrolyte membrane fuel cells, have a plastic electrolyte. The membrane material most widely used in PEMs is produced by DuPont and looks like the plastic wrap used for storing foods. The word "proton" refers to the hydrogen ion that passes through the polymer membrane.

PEMs are probably the best known of all fuel cell technologies because of their potential to replace the internal combustion engine in cars and trucks. Companies are also pursuing the development of PEMs in stationary applications to power homes, offices, and factories. Yet the very attribute that makes them attractive for transportation—operation at low temperatures (150°F)—makes them potentially less attractive than other fuel cells for powering a building or factory because their cogenerated heat has far fewer useful purposes (see Chapter 3).

To run on natural gas or other fuels, PEMs require an external reformer, a separate device to extract hydrogen from the fuel. This limits the overall electric efficiency of a PEM system running on natural gas to 35 to 40 percent, similar to that of current internal combustion systems. Moreover, they require very high-purity hydrogen and can tolerate only very low levels (in parts per million) of carbon monoxide before performance begins to decline significantly. Making a small, inexpensive reformer that produces such high-purity hydrogen from natural gas (or other fuels such as gasoline) has proven difficult. Research is being carried out on slightly higher-temperature PEMs, which would provide higher-quality heat and much greater carbon monoxide tolerance (thousands of parts per million).

As of 2003, commercial PEM products with significant sales in the United States did not exist. For powering a building, equipment prices below $4,000/kW will make PEMs competitive with existing phosphoric acid and molten carbonate fuel cells, though a price below $1,500/kW is probably needed to enable them to enter the market without a government subsidy, and that appears to be several years away. Roger Saillant, chief executive officer (CEO) of Plug Power, one of the leading manufacturers of PEMs for stationary applications, anticipates being able to deliver a fully installed 5 kW PEM system with reformer for less than $1,500/kW in the 2007–2008 time frame. For use in cars, a far lower price is needed. The goal of the DOE's program for PEM fuel cells is $45/kW. This will require major technology breakthroughs in the coming decade, as well as mass production—hundreds of thousands, if not millions, of units per year.

The pot of gold at the end of the rainbow for low-temperature fuel cells is to replace the internal combustion engines in hundreds of millions of cars, and that creates technical, cost, and logistical issues that are unique to PEMs. For instance, can hydrogen fuel be delivered to millions of PEM vehicles safely, conveniently, and affordably? Who will pay for this hydrogen fueling infrastructure, which may cost hundreds of billions of dollars? Such questions are addressed in Chapters 4, 5, and 6.

Other fuel cell types are the subject of continued research. For instance, *alkaline fuel cells,* with an aqueous solution of potassium hydroxide as an electrolyte, power the space shuttle, providing not only electricity but also potable water. These cells are very efficient but currently require exceedingly pure hydrogen and oxygen, since carbon dioxide (CO_2) in the air can poison the chemical reaction. *Direct methanol fuel cells,* which use a polymer electrolyte membrane (though other electrolytes are being explored), are also of great interest because they would transform methanol into electricity and heat without requiring a reformer. Currently, they are very expensive and have low power density (power produced per unit volume of fuel cell), among other problems, but they could be critical should the nation decide to pursue a "methanol economy" (see Chapter 4). On a smaller scale, these fuel cells can serve as energy sources for laptop computers and other portable electronic devices. Such micro–fuel cells, running on replaceable methanol cartridges, could well achieve significant commercial sales, but they would not have a significant effect on the energy and environmental problems that are driving the interest in larger fuel cells and the hydrogen economy, which are the focus of this book.

Reversible fuel cells are also the focus of much research. A fuel cell can be run "backward" to take an external source of electricity and turn it into hydrogen, which can then be stored and later used by the fuel cell to generate electricity. If it turns out to be more cost-effective than other means of storing electricity, such as batteries, this strategy could have a variety of potential applications in military and space missions. Reversible fuel cells could also help address one of the barriers to more widespread use of renewable electricity—its intermittency. For instance, excess electricity generated by windmills during windy times could be stored and used during less windy times.

Most of the initial R&D on reversible fuel cells has gone into PEMs, but the research suggests they are not ideal for this purpose. Half or more of the electricity is lost in the round-trip process of converting electricity to hydrogen and then back into electricity. SOFCs appear more promising, with the possibility of a round-trip efficiency of 70 percent or more. Remote areas may well represent a genuine niche market for reversible fuel cells, should they prove practical. But in the United States as a whole, we currently use so little intermittent renewable energy, such as wind power, that significant storage is unlikely to be required for decades. An analysis in the May 2003 issue of *Windpower Monthly* goes further, arguing that in the specific case of wind power, such expensive storage may never be either cost-effective or environmentally desirable (see Chapter 8).[9]

Promise versus Performance: A Caveat

Many novel energy technologies have failed to live up to their promise. Plug Power's CEO, Roger Saillant, said in June 2003 that "too many people have hurt the industry by overpromising and underdelivering." Andrew Skok, director of distribution support for FuelCell Energy, told me, "In the R&D phase, you tend to over-state what you can do. In the commercial phase, you understate what you can do." For that reason, Skok said the company's 250 kW unit would have an overall cogeneration efficiency of 70 to 75 percent, though he believed it would probably perform better. Commercial customers want warranties, and you don't want to disappoint them by underperforming.

These words are worth bearing in mind throughout this book. We have very little experience with commercial fuel cells, and the only one we do have experience with provides a cautionary tale. In the mid-1990s, the manufacturer of the PC25 promised an electric efficiency of 40 percent, sold the system for $3,000/kW (probably at a loss), and projected a price of $1,500/kW by 2000 as a result of further system optimization and manufacturing economies of scale.[10] By 2002, the system efficiency, as measured in more than a dozen military base installations over hundreds of thousands of hours, was much closer to 35 percent; the company had raised the price to $4,500/kW; and it was phasing out the product.

I recall a 2003 meeting between a fuel cell company CEO and a senior DOE official with three decades of government experience in fuel cells. The official asked the CEO when he would have a commercial product, and the reply was "two to three years." The official smiled and said he couldn't remember how many times in the past thirty years someone had made exactly the same claim to him.

Are current fuel cell cost targets realistic? The DOE goal of $400/kW for a complete fuel cell system would by itself just about guarantee that solid oxide fuel cells achieve major commercial success. But it is very ambitious, considering that we are still at ten times that cost after decades of R&D. The goal of $45/kW for PEMs is an even more ambitious target. To come within a factor of five of this target even with very high-volume manufacturing will require the solution of extremely difficult R&D problems such as reducing the quantity of platinum catalyst substantially or developing a non-platinum catalyst, achieving high degrees of stability of fuel cell electrodes and electrolytes (including overcoming sensitivity to impurities), and producing durable membranes that deliver high-performance and efficiency at very low cost.[11] Many experts I spoke to in and out of government believe the PEM cost goal may be impossible.

Most fuel cell companies are still in the R&D stage but are making very optimistic projections about performance and cost. It is possible they will achieve their goals. But it is premature and unwise to base major investments or policy decisions on such expectations. Successful performance in a laboratory setting is no guarantee of successful performance in a real-world setting. And both are quite separate from whether one can achieve the sales volume and economies of scale needed to achieve cost goals. The real-world problems and prospects for fuel cell commercialization are the subject of the next chapter.

CHAPTER 3

The Path to Fuel Cell Commercialization

Although fuel cells seem like a twenty-first-century marvel, they are a nineteenth-century invention that predates the invention of the internal combustion engine.[1] The internal combustion engine has become one of the most successful commercial products of all time. Fuel cells have largely failed in the marketplace, even after intense research and development (R&D) and commercialization efforts in recent decades. Why should we believe that marketplace success is now just around the corner?

There are several reasons. Advanced manufacturing techniques and miniaturization technologies, many derived from the semiconductor and computer industries, are now being applied to fuel cells. Manufacturers of the low-temperature fuel cells most suitable for transportation have significantly reduced their use of the pricey platinum catalyst. And the scale of corporate and government effort is greater than it has ever been, with many of the world's largest and best-known companies funding major R&D and commercialization efforts. Analysis suggests that fuel cells, especially high-temperature ones, can reach the low cost and high performance needed to succeed commercially in providing stationary power for buildings and factories—if they can achieve significant production volume to attain manufacturing economies of scale.

There's the rub. It is much like the chicken-and-egg problem of trying to create a hydrogen fueling infrastructure while selling a significant quantity of hydrogen fuel cell vehicles, as discussed in Chapter 1. The dilemma is this: Who will buy highly expensive fuel cells before they come down in price dramatically, and how will they come down in price without a great many people buying them? The problem is a serious one, since the only commercially available fuel cell of the past decade, the phosphoric acid fuel cell, failed in large part because it never achieved the kind of steadily growing sales needed to justify continuing R&D into improving technology and expanding manufacturing capacity—both of which are crucial for the cost reductions needed to make any successful breakthrough product.

The commercialization dilemma is especially acute in the United States, for three reasons. First, we have generally had among the least expensive electricity in the developed world.[2] Japan's electricity prices are more than double ours. Also, Japan's and Europe's markets have had substantially larger government subsidies for deploying clean energy technologies. Thus, alternative sources of power will be competitive sooner *outside* the United States. In the case of solar photovoltaics (electric power from sunlight), throughout the past decade the bulk of U.S. production was sold overseas— in Europe, Japan, and the developing world.

Second, we have the least expensive gasoline in the developed world. Gasoline taxes in Europe and Japan are several times higher than those in the United States, and, overall, gasoline prices there are two to three times higher than ours. Gasoline in Japan, France, the United Kingdom, and Germany costs between $3.60 and $4.60 per gallon.[3] Thus, again, fuel-efficient cars and alternative fuel vehicles will be competitive sooner *outside* our country. This is one reason why diesel engines, which are highly fuel efficient, are much more popular in Europe and constitute more than 40 percent of light-duty vehicles sold there, in contrast to only a small percentage here (see Chapter 8).

Third, although Europe and Japan are adopting strategies to reduce greenhouse gas emissions, we in the United States are not.

Many European countries are putting in place carbon dioxide (CO_2) taxes or trading systems whereby credit or cash would be given for reducing CO_2 emissions. This will further increase the competitive advantage of technologies that reduce greenhouse gas emissions, such as fuel cells that are run on fuels derived from renewable sources or that achieve high efficiency by simultaneously generating electricity and heat.

It would be unusual for a country to become a leader in a new technology without a very robust domestic market for that technology.[4] Hence, if the United States is to avoid having other countries become leaders in patenting, manufacturing, and deploying fuel cells and hydrogen technology, we must take strong measures to encourage domestic commercialization. As the U.S. Department of Energy's (DOE's) 2003 *Fuel Cell Report to Congress* concluded, "market forces alone are unlikely to result in large-scale use of fuel cells in the next few decades."[5] I have spoken to a number of early corporate adopters of multiple stationary fuel cell systems, and while they are attracted to the potential environmental and reliability benefits of having power generated on-site, they have typically demanded a fast economic payback. Historically, this has primarily come from government subsidies or massive price discounts by the supplier.

The rest of this chapter covers some near-term commercialization prospects and problems for stationary fuel cells. The chapter has two main points. First, if fuel cells can hit their long-term price and performance goals, they will do exceedingly well in the large and growing market for providing power to buildings and factories. Second, many of the early niche markets that companies have been betting on to accelerate commercialization and cost reductions are not as attractive as widely believed. Let me start with this second point.

The Market for High-Reliability Distributed Generation

Providing *highly reliable* power to businesses is commonly held out as a near-term commercialization strategy for fuel cells. The Royal

Dutch/Shell Group, in one of its two 2001 scenarios that gaze out fifty years into the future (see Chapter 7), has fuel cells play a prominent role, beginning with this very market: "Fuel cell sales start with stationary applications to businesses willing to pay a premium to ensure highly reliable power without voltage fluctuations or outages."[6] Every fuel cell business plan I have seen uses the reliable-power market as one of its entry points. Yet, as we will come to understand, this may not be an easy market to crack. *The projections that this market is both large and fast-growing have been built around a myth.*

A new energy technology (like any other new technology) seeking to break into a market already in the grasp of another, less expensive, technology will look for market niches in which its unique attributes provide a competitive advantage. For instance, photovoltaic power works without an external fuel source other than sunlight and without being hooked up to the electric power grid, so it can provide power in places that are far away from existing power lines. That is why solar power is so popular in the developing world, since there is no need to build either large, polluting power plants or thousands of miles of power lines and the accompanying infrastructure. In many developing countries, as well as for a number of remote applications in the developed countries, photovoltaics provide electricity at competitive prices.

A major reason why fuel cells (as well as photovoltaics and other clean, small-scale power sources) have received so much recent interest is our uneasiness over the reliability of the U.S. electric system. In the summer of 1999, major power failures hit Illinois, New England, New York, New Jersey, Delaware, Texas, and Louisiana. Other areas, including Colorado and the upper Midwest, narrowly averted major outages, often at the cost of thousands of lost work hours on the part of companies voluntarily curtailing operations — either to avoid high wholesale market prices or because they have "interruptible load" contracts with their utility. The summer of 2000 brought a repeat, with power failures hitting San Francisco, Detroit, and New York and warnings of outages throughout the country. By January 2001, voluntary measures in California had proved inadequate, and businesses and consumers throughout the state were subject to rolling blackouts. Utility deregulation in California, as the world now knows, proved a failure and subjected the state not only to unreliable power but also to extreme price shocks. The massive blackout that hit the United States and Canada in August 2003 underscored the inadequacy of our electric system.

The reliability of the electric power grid is also a growing worry because of the high-tech nature of today's workplace. In many parts of the country, the utility grid can deliver high-quality power—steady enough that it will not interfere with high-tech equipment—but only 99.9 percent of the time. That may be fine for households or the corner grocery, but it can be devastating for manufacturers, data handlers, or any other high-tech operation. Unreliable power now costs businesses and consumers tens of billions of dollars per year.

Fuel cells have three marvelous attributes that make them potentially very attractive for providing reliable power. First, they have no moving parts and thus hold the prospect of continuous, reliable operation over extended periods of time with very low operations and maintenance costs. Second, running on hydrogen or natural gas (or many other fuels), they emit negligible amounts of the urban air pollutants regulated by the U.S. Environmental Protection Agency (EPA) and by cities such as Los Angeles—important clean air standards that make it expensive and sometimes impossible to put traditional fossil fuel power plants in urban areas. So, rather than reliance on large, distant, unreliable central-station power plants, fuel cells afford the possibility of on-site generation right where the power is needed. Other distributed generation technologies, such as photovoltaics, can do the same.

Third, because stationary fuel cells can simultaneously generate both usable power and heat, they can achieve much higher overall efficiency when used on-site. Right now, large central-station power plants burning fossil fuels, the source of most U.S. electricity, are on average relatively inefficient, converting less than one-third of the energy in fossil fuels into electricity. The waste heat generated by that combustion is literally thrown away, and further energy is lost in transmitting the electricity from the power plant to the factory or building. *The total energy "wasted" by U.S. electric power generators equals all the energy that Japan uses for all purposes: buildings, industry, and transportation.* To provide heat, hot water, and steam, more fossil fuels are then burned inside our buildings and factories. The average building boiler converts only two-thirds of the fossil fuels to useful heat or steam (although new ones are often more than 80 percent efficient). By generating electricity *and* capturing the waste heat in a cogeneration system, much energy and pollution can be saved. Overall fuel cell system efficiencies can exceed 80 percent.

These advantages would seem ideal for an entry into the market for highly reliable power. As we saw in the introduction to this book, the 200,000-square-foot Technology Center of the First National Bank of Omaha installed a fuel cell system in 1999 for just that purpose, since a power failure can cost just one of its major retail clients $6 million per hour in lost orders.[7] The bank installed a system developed by the SurePower Corporation of Danbury, Connecticut, because it was the most reliable electric power source it could find. By maximizing the availability of the computer system to protect existing clients and attract new ones, the bank sought "an edge over its competitors," according to Dennis Hughes, the bank's lead property manager.

"The average uninterruptible power supply (UPS)/generator/ utility system has an availability of about 3-nines," explained Thomas Ditoro, the project's electrical engineer. The statistical term "availability" is industry jargon meaning the fraction of time the system is available for use.[8] A system available 99.9 percent of the time is a 3-nines system, a system with 99.9999 percent availability is a 6-nines system, and so forth. "The most redundant UPS can achieve an availability of 4-nines. The fuel cell system at First National Bank of Omaha has a calculated availability of 6- to 7-nines."

The difference is striking. With only 4-nines of availability, a high-tech center would have a 63 percent probability of at least one major failure over its 20-year life. A 6-nines systems has only a 1 percent probability of at least one major failure in 20 years. This could be the difference between business success and bankruptcy.

After it was installed in mid-1999, the system was independently verified to be delivering availability in excess of 6-nines. Even though the facility experienced some loss of power from the electric grid a dozen times as of mid-2003, the "system never missed a beat," according to Hughes. The bank uses this remarkable reliability as a lead feature in its marketing campaign and has increased its market share as a result. Hughes has said that the system's high reliability is a "competitive advantage. With SurePower, First National can raise our customers' service expectations while generating higher revenues."

While the initial cost of the fuel cell system was higher than that of the traditional UPS system, "the life-cycle costs, however, prove the fuel cell to be less expensive than UPSs," Ditoro notes, which is "remarkable, since electricity costs in Omaha are some of the lowest in the nation." The system SurePower designed for the bank uses four phosphoric acid fuel cells made by UTC Fuel Cells to provide 320 kilowatts (kW) of critical power. The fuel cells are combined with flywheels, advanced electronics, and other components to eliminate the possibility of disruptions. Since each of the four fuel cells provides 200 kW, the system has redundancy built in, which is part of the reason why it achieves such high availability. This also means that the system generates significant excess electricity—as well as free cogenerated heat—all of which it provides to the bank. That is what gives it a lower life-cycle cost. Moreover, the fuel cell does not require air-conditioned space, whereas a traditional UPS system would have required $28,000 in annual cooling costs— more wasted energy.

Compared with a traditional system using a UPS and the electric grid, the SurePower system had more than 40 percent lower emissions of CO_2 and less than one one-thousandth the emissions of other air pollutants. In 2002, Chris Robertson, an energy consultant based in Portland, Oregon, and I worked with the company to place a bid with The Climate Trust to sell the CO_2 "offsets" from the next SurePower project. The Trust buys CO_2 from selected projects that reduce emissions as part of Oregon's efforts to offset the greenhouse gas emissions of new power plants built in the state. In September 2002, The Climate Trust announced that the SurePower proposal was one of its winners and that The Trust had committed $1.2 million toward the purchase of those offsets. Thus, not only is the system capable of producing clean, inexpensive, highly reliable power; it can also provide a new revenue stream to early adopters.

And yet, for more than four years after its very successful First National Bank of Omaha project, SurePower was not able to sell a second high-reliability system. The interrelated reasons for this go to the heart of the reason why fuel cells will have difficulty using reliability problems as a path to commercialization.

Price and payback trump almost everything else. Those who construct new industrial and commercial facilities are driven by the desire to have the lowest initial capital cost—not the lowest life-cycle cost. In retrofitting or upgrading existing buildings and factories, most companies are driven by the need for a very rapid payback. Since the early 1990s, I have talked to and worked with dozens of energy managers and other corporate executives interested in deploying a variety of distributed generation technologies (including a few who have explored the possibility of a SurePower system). The main conclusion I draw is that while companies *say* they would like more reliable power, the vast majority are not willing to pay very much more for it, which makes it a tough market for fuel cell companies looking for customers willing to pay a premium for the technology.

A related issue is that it is much easier to make the economics work for on-site cogeneration in a new facility than in a retrofit. That is because in an existing facility the company has already paid for the capital cost of the heating and cooling system (and related infrastructure), as well as whatever systems the company employs for reliable power, typically a UPS system and backup diesel generators. To be cost-effective in a retrofit, a new technology must compete against the sunk cost of all that old equipment, no easy task for a new technology trying to break into the market with a relatively high price. In new construction, a fuel cell can compete directly with the alternatives on equal terms. Moreover, a new facility's energy system can be designed to utilize the waste heat effectively, thereby capturing more of the efficiency a fuel cell can deliver. This is often difficult in a retrofit. High-temperature fuel cells can run an absorption chiller, which turns heat into cooled water for use in air-conditioning, but these systems are expensive and are far more cost-effective in a new building than when replacing an existing air-conditioning system.

The market for new buildings and factories that require high-reliability power systems is a small fraction of the market for upgrading or replacing the equipment in existing facilities. So, to the extent a technology is largely limited to new construction, it will have a much smaller and more competitive potential market.

Another setback for manufacturers of fuel cells trying to break into the distributed generation market has been the tremendous volatility of natural gas prices. In just the five-year period from the beginning of 1999 to the end of 2003, we've had two major spikes in the price of natural gas, during which time companies that have dual-fuel facilities have shifted from natural gas to oil. Since most stationary fuel cells run on hydrogen produced from natural gas, the volatility in the price of gas translates into volatility in the price of power from the fuel cell. Most companies are risk averse and very nervous about committing to a power source that is dependent on a fuel whose price can double in a short period of time. This is a particular problem with low-temperature fuel cells, which have less fuel flexibility and less efficiency.

The promise of a huge high-reliability market was built around a myth. Understanding the true potential size of the high-reliability market requires an understanding of the complex interplay between the growth of the Internet and the digital economy's effect on electricity demand. Early analyses of these issues created a myth that raised false expectations in the market.

In May 1999, *Forbes* magazine published an article titled "Dig More Coal—the PCs Are Coming," by Peter Huber and Mark Mills. The article claimed that in 1998 the Internet consumed 8 percent of U.S. electricity, up from less than 1 percent in 1993, and computers and other office equipment consumed 13 percent of the country's electricity, up from 5 percent in 1993. Because the Internet is growing explosively, the article predicted, "it's now reasonable to project that half of the electric grid will be powering the digital-Internet economy within the next decade."[9]

These statistics then cropped up in many reports by financial institutions and presentations by major energy companies, the business plans of most distributed energy companies, explanations for power reliability problems in California, and myriad newspaper and magazine articles, including articles in *Fortune* magazine, the *Wall Street Journal,* and *Newsweek.* Mills and Huber began publishing an investment advisory newsletter called *Digital Power Report,* recommending stocks of a variety of distributed energy companies, including fuel cell companies, on the basis of their projection of an explosive growth in the need for "digital power." Many brokerage firms followed suit, stoking the fires with promises of explosive growth and helping to create a volcanic eruption in the stock prices of such companies as Capstone Turbine Corporation and FuelCell Energy Inc.[10]

There is just one problem with the analysis: *All the numbers are wrong.* Since the myth persists throughout the media and in the business and strategic plans of hydrogen and fuel cell companies, it deserves reexamination.

Almost immediately after Mills and Huber published their analysis, scientists at Lawrence Berkeley National Laboratory examined the underlying numbers in detail and found that the estimates of electricity used by the Internet were *too high by a factor of eight.*[11] Major overestimates were found in every category, including the authors' calculations of energy used by major dot-com companies, by the country's Web servers, by telephone companies' central offices, by Internet routers and local networks, and by personal computers used in business and at home. The Internet did not consume 8 percent of U.S. electricity in 1998—it was closer to 1 percent. Computers, office equipment, and semiconductor manufacturers do not consume 13 percent of electricity—it is only 3 to 4 percent.

Also, growth in electricity consumption has not increased since the surge in the Internet—in spite of higher growth in gross domestic product since 1995, hotter summers, and less demand-side management funding by electric utilities, all of which should contribute to *higher* electricity growth.[12] The Internet may actually be slowing the growth of electricity consumption by making the overall economy operate more efficiently and productively.[13]

A 2002 study by Arthur D. Little Inc. on the energy consumption of office and telecommunications equipment came to a conclusion similar to that of Lawrence Berkeley National Laboratory.[14] So did a 2002 study by the Rand Corporation.[15] I am unaware of a single independent study that supports the original Mills-Huber analysis.

In a December 2002 *Wall Street Journal* article, "Bold Estimate of Web's Thirst for Electricity Seems All Wet," David Wessel reviewed the matter and commented, "It's an instructive case study in how a simple assertion of questionable accuracy can be repeated so often that it is accepted as fact by people in the press, in government and on Wall Street who ought to be more discerning."[16] This is yet one more lesson that major investment decisions by governments, companies, and individuals should be made using conservative assumptions.

What of the future? After the collapse of the Nasdaq in 2000, the telecommunications sector fell into a prolonged slump. It became much harder to raise money for new companies and new projects in the public and private capital markets. Virtually all the companies commercializing products for this sector, including those developing high-reliability and distributed energy technologies, also went into a prolonged slump.

Even after the slump ends, however, the market is unlikely ever to come close to its pre-2000 expectations because the premises of the boom were flawed. Fundamentally, the Internet does not consume a large fraction of U.S. electricity today, nor do personal computers, office equipment, and network equipment; furthermore, the growth rate of power used by the Internet is much slower than the growth rate of the Internet itself.[17]

Hence, while the digital power market is important, it is not as large and fast-growing as once thought. Not an ideal place for costly emerging products.

There is more to power availability than meets the eye. A small percentage of electricity in the United States is devoted to corporate computers, information technology, and the Internet, but it does not follow that more than a small percent of companies with computers need the kind of high-availability power that might come from fuel cells. Consider your own company. How often is the power out? Rarely. How often does the network go down? Quite often. Yet your company is still in business. Clearly, the power system is much more reliable than your computer network system, and the relatively unreliable network is not fatal for the vast majority of companies. Why would your company pay a premium to make its power system more reliable, at least until its computer network is much more reliable?

A key distinction must also be drawn between power quality and power reliability. The electric system typically has a great many voltage surges and other very short disruptions that can interfere with computer equipment. But for the vast majority of companies, that problem can be addressed with power electronics and UPS systems to smooth out the bumps. The more you are worried about power quality, the more sophisticated a system you can buy. Having the power go out entirely is a relatively rare event—but as we all saw with the massive blackout of August 2003, rare events do happen.

The blackout may spur businesses to reconsider the value of having on-site generation, although, if they do, the beneficiary may not be fuel cells, since only a very few are starting to be commercialized. For the time being, the competition, such as gas turbines or reciprocating engines, remains much cheaper, as discussed later in this chapter. On the other hand, the blackout should motivate the federal government to make immediate investments in our nation's electricity infrastructure, including the aging transmission and distribution system; in improved software and hardware to monitor the grid and permit diagnosis before problems get out of hand; and in energy efficiency to reduce the demand for peak power that can destabilize the grid. This would have the effect of making the grid much more reliable.

Facilities in which reliable power is a concern today, such as hospitals, typically have one or more backup diesel generators. This is hardly an ideal solution, since those generators are too polluting to operate very often in most major cities, so the facilities have paid for a piece of hardware that they use only in emergencies. It is, however, a relatively cheap and dependable solution for most applications.

Moreover, contrary to the claims of many business plans I have read, you cannot deliver power that is significantly more reliable than traditional backup systems by simply building a few fuel cells. This is another grand myth. Designing a high-availability system requires use of the well-established science of probabilistic risk assessment (PRA). The key design element is to avoid so-called single points of failure. A single point of failure in a system is an individual component that, if it fails, brings down the whole system.[18] Simply stringing together diesel generators or fuel cells cannot provide high availability. Redundancy of all components is required, as well as a sophisticated electric system that itself has built-in redundancies and fail-safes.

These requirements pose two serious problems for the scenario in which fuel cells become commercially successful by using the high-availability market for early entry. First, the need for redundancy penalizes on-site generation devices that are expensive to begin with relative to the competition by further driving up the total system cost, especially for facilities that need a considerable amount (i.e., several megawatts) of on-site energy, such as semiconductor manufacturing plants. This is one reason why Sure-Power's second-generation design (for multi-megawatt on-site generation) dropped the phosphoric acid fuel cells and integrated new, less expensive, low-polluting power units.[19]

Second, guaranteeing availability as high as 99.9999 percent requires a tremendous amount of performance data on every single piece of equipment. Accurate risk-assessment analysis requires reliable data such as "mean time between failure" (how long a component is likely to run before breaking down) and "mean time to repair" (how long it will take to fix a component that has broken down). Analysts would prefer to have as much as 1 million hours of data on each and every system component. That takes years to gather for any new technology. And if a fuel cell company is still improving its product and keeps releasing new models, then data on the previous system may not be useful. As Paul Lancaster, vice president of finance for Ballard Power Systems Inc., explained in October 2002, for a product to truly be considered commercial, it must have a "frozen design" upon entering the market and "specifications and attributes that customers can be assured will be met in the product's performance."[20]

My point is *not* that fuel cells cannot provide high-availability power—they can—but only that this market is particularly difficult for product entry. In fact, from the customer's side, the more a company cares about high-reliability power—the more a power outage would hurt its business—the more risk averse it is likely to be. Companies willing to pay a premium for such power are exactly those least willing to "bet their company" on a new technology. They can't take any chances, not even small ones. The hype that surrounded distributed generation at the turn of the twenty-first century, and the resulting disappointment in the performance, cost, and availability of a variety of technologies, including microturbines and fuel cells, has further served to create the perception that these technologies are not yet ready for prime time. The bottom line:

- The digital power or premium power market is much smaller and is growing much more slowly than has been widely touted.
- A significant number of companies that need premium power have a system that works well for them (or at least they think it does) and won't pay much more to obtain a new system or replace an existing system.
- New technology needs time to establish itself.

In the long run, the market for stationary fuel cells will no doubt benefit from their potential to deliver more reliable power than traditional power plants. But trying to capture customers who need high-availability power is probably not a strategy that will dramatically accelerate market entry for fuel cells in the near term, when they are still relatively pricey and relatively untested compared with the competition.

The Residential Market

Another market widely mentioned as promising for early deployment of fuel cells is provision of power to homes. A diverse group of analyses—by the Department of Energy, Princeton University, Directed Technologies Inc., and the Capital E Group—suggest that homes are not a very attractive market for fuel cells, at least in the United States for the foreseeable future.[21] Nor would residential fuel cells have a significant positive effect on greenhouse gas emissions. The home market is worth examining here not merely because so many fuel cell companies are pursuing it, especially manufacturers of proton exchange membrane (PEM) fuel cells, but also because so many companies and analysts have put forward the possibility that residential fuel cells could be part of an overall home hydrogen fueling system for fuel cell cars (see Chapter 6).

The average American home consumes about 1 kW of power over the course of the day, with a peak of slightly more than 4 kW. The electricity is used for the refrigerator, lighting, home electronics, and other appliances. The electricity load is highest in the summertime, when air-conditioning is being run. Heat and hot water are typically provided by natural gas, although fuel oil and electricity are not uncommon. The hot-water demand is relatively constant year-round (and somewhat smaller than the power load). Heating demand peaks in the wintertime, and it is far greater than the electric power draw at that time.

This means that a relatively small fuel cell, about 1 kW in capacity, can cogenerate year-round, providing base-load power and hot water. Larger fuel cells will provide far more heat than a home can use most of the time, and they will end up throwing heat away, reducing their overall efficiency, cost-effectiveness, and environmental benefit. Not surprisingly, the DOE's analysis found that the optimal size for a residential combined heat and power (CHP) or cogeneration system in the United States was 0.73 kW for a PEM fuel cell.[22] That is significant for two reasons.

First, small residential PEM fuel cells will be perhaps 35 percent efficient or less in converting natural gas to electricity. When we average out the CO_2 emissions from all the U.S. power plants, from coal to nuclear to hydroelectric, we see that the entire U.S. electric grid is roughly the equivalent of one enormous 30 percent efficient natural gas power plant. So, *unless the home* PEM *is cogenerating usable heat most of the time that it is operating, and thus replacing the natural gas (or other fuel) used to produce hot water, the system is not avoiding significant greenhouse gas emissions.* The DOE's analysis found that the home PEM system would cut CO_2 emissions by less than 10 percent. A larger PEM would achieve even smaller savings, since a smaller fraction of the waste heat would be utilized. And, if we view the home PEM as a new power plant, it has even tougher competition in the battle to avoid global warming: Most of the new central-station power coming online in the first decade of this century is from natural gas combined cycle plants, which have an efficiency of 55 percent or more.

Second, the very small residential PEMs that might make sense from an environmental perspective are the least likely to make sense from an economic perspective. As noted in the Introduction to this book, long before PEM fuel cells become inexpensive enough to compete with gasoline engines (which requires a price below $50 per kW), PEMs will become competitive with stationary power plants, which can cost $500/kW or more. One big difference for stationary PEMs, however, particularly in homes, is that for the foreseeable future they will need a reformer to extract hydrogen from natural gas in order for hydrogen to be convenient and cost-effective as a consumer product. As an aside, it would be a great waste of energy for homeowners to use their own electricity to electrolyze water to create hydrogen to run a fuel cell to generate their electricity; also, delivery of hydrogen to the home will very likely be prohibitively expensive and less environmentally beneficial, at least for the foreseeable future (see Chapters 4 and 5).

In any case, developing an affordable, small-scale reformer capable of delivering the kind of high-purity hydrogen that a PEM requires has proven very hard. Ballard, the leading PEM fuel cell company, is phasing out its fuel reformation business. In early 2003, the company began marketing a 1 kW PEM for about $6,000 that runs on pure hydrogen. The product will come with storage tanks for industrial and residential users.[23] In June 2003, Plug Power introduced a 5 kW PEM system to provide backup power in the telecommunications industry, at a price of $3,000/kW.[24] This system runs on hydrogen and is designed for some 1,500 hours of use over ten years. Another company quoted me a current price for a PEM plus reformer of $10,000–$12,000/kW.

Costs seem likely to decline rapidly, but as the Directed Technologies analysis in 2000 explained, "based on our evaluation, many of the fuel cell and fuel processor costs are fixed, independent of power output. As the systems become smaller, the cost per kilowatt increases." A 20 kW unit might ultimately cost $1,500/kW, while a 2 kW unit might cost more than $5,000/kW. Also, installation costs are likely to be high in the beginning, requiring both a specially trained plumber and a trained electrician, since the fuel cell will have to be integrated with the home's electric and hot-water systems and possibly its heating system. The analysis concluded, "In order to bring a 10% real, after-tax return on investment, a company would still have to charge a single-family residence about 40 cents/kWh, well above the average residential rate near 8 c/kWh." Analysts at Princeton University's Center for Energy and Environmental Studies came to a similar conclusion: "Single-family PEM fuel cell units do not appear to be economically viable." And this was using relatively optimistic assumptions about fuel cell costs—an installed cost of less than $1,000/kW at high-volume production.[25]

Another optimistic assumption is that the PEM system will be used with "net metering": When the system is generating more electricity than the home is using, power is returned to the electric grid and the electric meter runs backward, every homeowner's hope. As of 2003, about half of the states had net metering for residential power systems such as photovoltaics and fuel cells.

Unfortunately, though, consumers are typically allowed to sell electricity back to the utility at their relatively high retail rates only for electricity production that does not exceed their total *annual* consumption. For any excess production, a homeowner receives the rate that a utility pays to a large central power plant (the utility's so-called avoided cost). This rate is far lower than the average residential rate. So if your home consumed 10,000 kWh of electricity during the year, but your fuel cells generated 30,000 kWh, you would typically get the retail price for only the first 10,000 kWh generated, but a much lower price for the remaining 20,000 kWh. Utilities would lose a considerable amount of money if they had to purchase electricity from millions of small residential fuel cells at the retail price, especially overnight, when they have a lot of excess capacity and prices are very low, which is precisely the time that a home's power consumption is also lowest and its fuel cell would have excess capacity.

Yet, as noted earlier, large home PEM fuel cells will probably not provide much net savings of energy or much net environmental benefit, if any, at least in regard to greenhouse gas emissions. Princeton researcher Tom Kreutz has said, "I strongly doubt that the advantages of net metering will be available to more than a tiny fraction of power generation capacity, and will be reserved primarily for encouraging renewables (as it should be, to my mind)."[26]

Another practical issue is this: Just as with large fuel cells for commercial buildings, it is much easier to make the economics work for on-site cogeneration in a new home than in a retrofit. Again, in an existing home, the consumer has already paid for the hot-water and heating systems, and a new technology must compete with the sunk cost of all of that equipment. A plumber and an electrician would have to be brought in to figure out how to make the fuel cell work in an existing home, assuming there is space available that meets whatever codes and standards are ultimately devised. Installation will be easier and faster in a home designed around a fuel cell. The market for putting fuel cells into new homes, though, is obviously far smaller than the potential market for putting them in existing homes.

The reliance of fuel cells on natural gas to produce hydrogen is also extremely problematic. In the first few years of the twenty-first century, homeowners have experienced the price shock of a surge in residential natural gas prices twice. How many who already use natural gas for home heating and hot water will want to undiversify their fuel mix and have their home electricity bill also entirely dependent on the price of gas, especially when the cost savings will be minimal (at market entry prices for PEMs), as will be the environmental benefits?

Again, this is not to say that no one will buy a home PEM. There are parts of the country where electricity prices are very high, where there are electricity transmission and distribution constraints, and where there are severe air quality problems that make it difficult to build new generation. This is a small but genuine market, especially for people with large homes. Also, first adopters of technology will no doubt be interested in home fuel cells. My point is merely that the market is far less attractive than is widely perceived, especially for a market entry of PEMs.

Both the 2000 DOE analysis and a 2003 analysis by the Fuel Cell and Hydrogen Research Centre in Berlin suggest that solid oxide fuel cells (SOFCs) may be a better candidate for home fuel cells because they have higher electric efficiency, they do not need an expensive external reformer, and they have more usable heat.[27] For home use, however, SOFCs would have their own limitations. Since they operate at very high temperatures, they take a long time (several hours) to warm up, which is why they operate much better in commercial and industrial applications that require high levels of electricity continuously.

The Market for Cool Power

Another niche for market entry of stationary fuel cells, both in residences and in other facilities, is created by the willingness of some consumers and businesses to pay more for products that are perceived as "green." This may be the best road to early commercial success.

Companies in the United States are increasingly concerned with the environmental effects of their energy consumption. In particular, more and more corporations are taking action to reduce their CO_2 emissions, both direct emissions from their on-site combustion of fossil fuels and indirect emissions from purchased electricity (that is, the CO_2 emitted by the power plants they buy power from). For instance, the World Wildlife Fund (WWF) and my non-profit organization, the Center for Energy and Climate Solutions, have been working with businesses such as IBM, Johnson & Johnson, and Nike in the Climate Savers Program to develop and adopt innovative climate and energy solutions. As part of the program, we help companies develop aggressive greenhouse gas reduction targets. By 2010, the combined commitments of the first six Climate Savers companies will result in annual emissions reductions equivalent to more than 12 million metric tons of CO_2.

As discussed in my 1999 book *Cool Companies: How the Best Businesses Boost Profits and Productivity by Cutting Greenhouse Gas Emissions*, there are two basic strategies for reducing emissions: Use energy more efficiently, by doing things such as choosing a more efficient light bulb, or use what I call "cool power," sources of energy that generate lower amounts of greenhouse gases, such as renewable energy sources or cogeneration. In some—but not all—circumstances, stationary fuel cells can help a company reduce CO_2 emissions.

Not surprisingly, then, many of the companies that are purchasing fuel cells today, or serving as deployment sites for precommercial units, are doing so for environmental reasons. The federal government has for many years offered a $1,000/kW subsidy both to encourage market entry of fuel cells and to reduce pollution. A number of states, such as New Jersey, Connecticut, and California, have begun offering subsidies, some in excess of $2,000/kW, to help buy down the cost of fuel cell purchases. The green market, supported by these subsidies, may be the most important path to commercialization for stationary fuel cells.

For instance, in 2002, Starwood Hotels and Resorts Worldwide Inc. announced it was putting two molten carbonate fuel cells into its Sheraton Parsippany and Sheraton Edison hotels in New Jersey.[28] Starwood is taking advantage of a $1.6 million grant given by the New Jersey Clean Energy Program to support fuel cell purchases. The two 250 kW fuel cells from FuelCell Energy each weigh about forty tons and are roughly the size of a railroad car.

Why is Starwood doing this? The company is "getting a cost benefit, and it is clean, green energy," according to John Lembo, director of energy for Starwood. The fuel cells will cogenerate, providing about one-quarter of the electricity and one-quarter of the hot water needed for each hotel.

The fuel cell installation was done as an outsource deal in that Starwood does not own or operate the fuel cells—the PPL Corporation, a Pennsylvania-based utility, does. This eliminates all the hassle and up-front capital costs, which would otherwise have made this far less attractive to Starwood. The hotel company is not in the power generation business, nor does it want to be. It put up only some $40,000. PPL guaranteed a lower electricity price than Starwood was currently paying. Lembo told me that this project was "not a huge economic savings, but it's as green as you can get in power generation in a hotel."

Starwood has a culture of energy efficiency. The company won the 2002 award for Excellence in Energy Management for the hospitality sector from the EPA's Energy Star program, which helps promote and recognize best practices in energy efficiency. In 2003, Starwood was named EPA's Energy Star Partner of the Year in the hospitality sector—a mark of the company's leadership role in energy efficiency. This is yet another reason why the company is installing two fuel cells. "We want to be on the cutting edge of this kind of technology," says Lembo. "This is the future."

Another early adopter of new fuel cell technology is the Dow Chemical Company.[29] In May 2003, Dow announced it would begin installing and testing 75 kW PEM fuel cells from the General Motors Corporation, which are being developed for automotive applications. Tests will take place in Freeport, Texas, at Dow's largest chemical-manufacturing plant. If the tests are successful, Dow would use as much as 35 megawatts (MW) of power from PEMS— the biggest fuel cell deal to date.

The GM fuel cells will not use a reformer but instead will run on pure hydrogen that is produced at the Freeport plant as a by-product of the chemical-manufacturing process. Some excess hydrogen is currently being combusted at the facility to provide power, and while hydrogen combustion can be more efficient than natural gas combustion, the hope is that the GM fuel cells will operate even more efficiently, with an electric efficiency of 45 percent or more. Also, Dow plans to use the cogenerated hot water from the PEMs, although the temperature of the water is not as high as Dow normally uses in its process. One possible application is to preheat the water going into the steam boilers.

A key reason why Dow pursued this deal is the environmental benefit. Dow has a long history of energy efficiency: The company already cogenerates more than 90 percent of the power it uses. As of mid-2003, Dow was calculating its overall greenhouse gas baseline and developing a greenhouse gas strategy. When I asked George Kehler, commercial manager for Dow's energy business in charge of green power market development, why Dow made this deal, he explained, "We cannot keep using energy the way we have." Reducing CO_2 emissions in particular was a motivation for Dow. Peter Molinaro, the company's vice president of government affairs, said at the May 2003 press conference announcing the deal, "If we are to achieve a more sustainable future and reduce society's footprint on the global climate, we will have to develop new technology, and raw materials."

Neither Dow nor GM has reported what the financial terms of the deal were. Dow insisted on a transaction that made economic sense for the company, and cogeneration is Dow's low-cost alternative to electricity from GM's fuel cells.[30] When PEMs finally become a commercial product, it is very unlikely that their price will be competitive with industrial cogeneration, but this deal still makes sense for GM as part of its overall strategy to commercialize its fuel cells and get real-world testing.

Emissions of High-Temperature versus Low-Temperature Fuel Cells

Because a great many governments, companies, and individuals will be considering fuel cells as part of their efforts to reduce their adverse environmental effects, it is worth examining the emissions benefits of stationary fuel cells in more detail.

Fuel cells are widely seen as an environmentally preferable way to deliver power. But this is true in an unqualified sense only in terms of emissions of urban air pollutants, such as particulates and oxides of nitrogen. As we have already seen in the residential discussion, this is not always true in the case of greenhouse gas pollution (i.e., CO_2), particularly for low-temperature fuel cells. As noted earlier, considering only CO_2 emissions, the U.S. electric grid is roughly the equivalent of one enormous 30 percent efficient natural gas power plant.[31] So we can see that the relatively modest reductions in greenhouse gas emissions that PEM fuel cells might provide a homeowner are also likely to be realized by the owner of a commercial building or factory using a PEM running on natural gas.

Fuel cells represent the purchase of a new technology for a company (or individual). To make a valid analysis, one should compare it with the alternative ways the purchaser could spend money to reduce emissions, such as buying a new boiler or a different cogeneration unit or renewable energy (or investing in energy-efficient lighting, for that matter).[32] It is easy to be confused when the efficiency of fuel cells is quoted as 80 percent or more (the total of the electricity and the heat that is provided) and then that number is compared with the average efficiency of the grid, 30 percent, suggesting that fuel cells are more than twice as efficient as the grid, which in turn leads to statements that emissions reductions will exceed 50 percent. That calculation, however, compares apples and oranges. The electricity from the fuel cell is properly compared with the electricity produced by the grid, which *is* generated inefficiently. But the *heat* from the fuel cell offsets, typically, the heat from a gas-fired boiler, which can be quite efficient. Commercial and industrial boilers can be 80 percent efficient or more, especially new ones.

This means that the higher the electric efficiency of a stationary fuel cell, the more primary energy and greenhouse gas emissions it is likely to save, since it is substituting for central-station power that is inefficiently produced. Also, as noted earlier, fuel cells running on fossil fuels that are primarily used to provide electricity and are not cogenerating most of the time may provide little or no CO_2 benefit. The less waste heat that can be used, the less CO_2 will be avoided.

That means that PEMs are at a double disadvantage from a greenhouse gas perspective. First, they have a relatively low electric efficiency. A one-year test completed March 2003 of a 3-kw stationary PEM fuel cell at a military facility measured an overall electrical efficiency for the system *running on hydrogen* of 27 percent.[33] While the technology continues to improve, residential PEMs of one to several kilowatts running on natural gas may ultimately be only 35 percent efficient; larger units for commercial buildings and industrial facilities could hit 40 percent. Second, their cogeneration capability is sharply limited because their waste heat is not hot enough for most industrial applications, which typically require steam, often high-pressure steam.

In particular, the exhaust temperature of PEMs is currently too low to run an absorption chiller, which uses heat to enable a refrigeration cycle. Commercial office buildings in most regions of the country typically do not require heating for most of the year, and they have only modest hot-water needs. A great many commercial office buildings, however, use air-conditioning not just in the summer but most of the year, as they must remove the heat generated by the people and by the lights, computers, and other electric equipment in the building.

Thus, PEMs will for the foreseeable future have a limited cogeneration capability in most commercial buildings and most industrial facilities. The Dow case was special because the Texas facility had excess hydrogen that was not being put to particularly good use, and PEMs may be able to generate electricity directly from hydrogen more efficiently than electricity can be generated by burning hydrogen—but this case is likely to be an exception. One could imagine running a stationary PEM on hydrogen generated from renewable electricity, but we are decades from having the kind of surplus renewable electricity that would justify throwing half of it away, merely to generate electricity on-site with a PEM (see Chapter 8).

Phosphoric acid fuel cells, which operate at a slightly higher temperature than PEMS, also offer only modest CO_2 savings when cogenerating, in most cases. The SurePower system at the First National Bank of Omaha, which used four PC25s, had more than 40 percent lower CO_2 emissions than the alternative system using the electric grid plus a UPS—but that was the result of several factors unique to this high-reliability application. Not only is the UPS itself directly less efficient than the SurePower system (that is, it requires more electric input to deliver the same amount of critical power), but it also requires a considerable amount of additional power for cooling. Those factors alone account for most of the CO_2 savings. In addition, the bank was able to use about one-quarter of the thermal energy provided by the fuel cells and achieved an overall efficiency of about 54 percent.[34]

High-temperature fuel cells, on the other hand, can offer substantial CO_2 savings. With electric efficiencies of 50 percent and high-quality heat, molten carbonate fuel cells and SOFCs hold the prospect of cutting CO_2 emissions in factories and buildings by one-third or more. Should hybrid fuel cell and gas turbine systems become economical, they could potentially reduce by half or more the CO_2 emissions of the systems they replace.

So, while it is true that customers who are environmentally responsible represent a significant market entry target for stationary fuel cell companies, the number of companies willing to pay a significant premium for big-ticket items such as fuel cells (and photovoltaics) is relatively small. In most cases, project economics and return on capital will have to be just about as good as other opportunities a company has. High-temperature fuel cells can bring substantial CO_2 savings; low-temperature ones do not when their fuel is natural gas, as it will be for most users for the foreseeable future.[35]

The Enormous On-Site Generation Market

If producers of stationary fuel cells can dramatically reduce their costs—by continuing R&D, obtaining government subsidies to spur commercialization, and finding early market niches—they will be able to compete directly in their ultimate market: small- to medium-sized power plants for on-site generation, especially cogeneration in factories and commercial buildings. Sustained success in this market will probably require a fully installed cost for fuel cells of $800–$1,300/kW.[36]

Many studies have shown that the potential market is enormous. For instance, a 2000 study for the DOE's Energy Information Administration found that the technical market potential for combined heat and power (CHP) at commercial and institutional facilities was 75,000 MW, of which more than 60 percent was in systems less than 1 MW in size. This sub-MW market is a very good match for fuel cell technologies. The remaining technical potential in the industrial sector is about 88,000 MW. That analysis did not look at the opportunity created by heat-driven chillers to expand the market for cogeneration, nor did it contemplate cost-effective systems in sizes below 100 kW, as are being pursued by a number of fuel-cell (and other) companies.[37]

Fuel cells will have two main difficulties in capturing a large share of the untapped CHP market. First, they have fierce competition in a number of very mature, reliable, low-cost technologies. Second, there are solid reasons why so much of this market is untapped: Many barriers exist to widespread use of small-scale CHP systems.

Let's start with the competition, since existing technologies and existing companies can be a formidable bulwark against new technologies and new companies. In fact, the reason renewable energy technologies do not have deeper market penetration in the United States today is *not* that those technologies failed to meet cost and performance goals; it is primarily that the competition did not sit still. It has proven much tougher, much more nimble and inventive, than had been expected, as discussed in Chapter 8. A variety of technologies have been pursuing on-site CHP for years, including gas turbines, reciprocating engines, and steam turbines.

Gas turbines (500 kW to 250 MW) generate electricity and heat in a thermodynamic cycle known as the Brayton cycle. They are a mature technology with an estimated 40,000 MW of installed capacity for CHP in the United States.[38] Electric efficiency for smaller units, less than 10 MW, is 25 to 30 percent, with overall efficiencies reaching 70 to 80 percent when the cogenerated heat is used. They generate relatively small amounts of nitrogen oxides (NO_x) and other pollutants, and a number of companies have developed ultra-low-NO_x units. Their high-temperature exhaust can be used to make industrial process steam and to run steam-driven chillers.

A typical 1 MW unit might cost $1,800/kW *installed*, and a 5 MW unit might cost $1,000/kW *installed*. It is difficult, however, to make a comparison between those numbers and the installed cost of fuel cells discussed in the previous chapter, since there are as yet no commercial fuel cell systems with high-volume sales. For both of these turbine systems, the turbine generator itself represents only about one-third of the total cost. Other major costs include the heat recovery steam generator, the electric equipment and cost of interconnecting to the grid, labor costs, and project management and financing.

Reciprocating engines are another very mature technology used for CHP.[39] They are the stationary version of automobile engines and come in two basic types: spark ignition Otto-cycle engines and compression ignition diesel-cycle engines. They range in capacity from a few kilowatts to more than 5 MW, and, running on natural gas, their electric efficiency can range from 28 percent for the smallest units to more than 40 percent for the largest ones. Capturing waste heat results in overall efficiencies of 70 to 80 percent. The high-temperature exhaust (700°F–1,000°F) can serve many industrial process needs and run an absorption chiller. Some 800 MW worth of stationary reciprocating engines are currently installed in the United States.

While the technology has seen steady development over the past 100 years as a result of its use in automobiles, there has been dramatic progress in the last three decades in increasing the electric efficiency and power density of reciprocating engines as well as reducing their emissions. A number of companies have been introducing units that can meet even California's strict urban air quality standards when running on natural gas, including the Hess Microgen unit, which can trigenerate electricity, absorption cooling, and hot water. The Starwood hotel chain, for instance, has begun installing Hess Microgen units in a number of its hotels.

A typical 100 kW reciprocating engine costs $1,500/kW fully installed, and an 800 kW unit might cost $1,000/kW. The engine itself accounts for only one-quarter of this price (or less for the smallest units). Again, the bulk of the price comes from the heat recovery system, the interconnect/electric system, labor and materials, project management, construction, and engineering costs.

Steam turbines are an even more mature technology, having provided power generation for more than 100 years.[40] Most central-station power is produced by steam turbines. A steam turbine does not directly convert fuel to electric energy; rather, it relies on a separate heat source, typically a boiler, which itself can run on a variety of fuels, such as coal, natural gas, petroleum products, uranium, wood, and various waste products such as wood chips or agricultural by-products. Steam turbines can be as small as 50 kW or as large as hundreds of megawatts. In 2000, some 19,000 MW worth of combination boiler and steam turbines were used to provide CHP in the United States.

For distributed generation, a boiler and steam turbine system can be expensive to build from scratch. But a company that already has a boiler providing high-pressure steam can often install a back-pressure steam turbine generator to provide low-cost and high-efficiency power generation. The technology works by taking advantage of pressure drops in existing steam distribution networks to generate electric power—using energy that is already available in the steam. From the point of view of cost and environmental performance, a back-pressure turbine is able to convert natural gas (or other boiler fuels) into electricity with an efficiency exceeding 80 percent, making it among the most efficient of all distributed generation devices. Not only are the CO_2 emissions low, but also the urban air pollution emissions are only that of the incremental fuel burned at the boiler, which are also typically quite low. In addition, the fully installed capital cost for these systems is typically in the $500/kW range. The combination of high efficiency, low cost, and low maintenance makes it a very cost-effective power source. Typical back-pressure installations have payback times of two or three years or less.

Sometimes the competition to a CHP project is not another technology, but a price break from the local utility or procrastination by the potential customer.

A lower utility bill is another very tough competitor. When the local utility learns that a company is considering cogeneration, it sometimes offers a lower electricity rate in return for an agreement *not* to cogenerate for a specified period of time. This is particularly true for bigger projects or those in which a company is contemplating replacing a large fraction of its total electric load with on-site generation. Such an offer can be very seductive, since it would save the company both the expense of making a major capital investment and the effort of figuring out how to integrate the heat and power into their processes. Most important, a lower utility bill reduces the future energy cost savings from the CHP project and thereby reduces the return on investment and pushes out the payback time. It is a not-so-hidden industry secret that some companies pursue CHP projects primarily as leverage to get a local utility to offer them a discounted price.[41]

The status quo is undeniably the biggest competitor to CHP projects. The easiest thing for a company contemplating a CHP project to do is to defer any decision indefinitely. Even without the local utility offering a discount on electricity, there are myriad other barriers to distributed energy projects that slow the process down, add cost, increase project complexity, and in general encourage inaction and kill projects. Sean Casten, chief executive officer of Turbosteam Corporation, which sells back-pressure turbine systems, put it best: "Our toughest competitor is 'Do Nothing.'"[42]

At the request of customers, vendors, and developers of distributed energy systems, the DOE launched a study of these barriers.

The result is a July 2000 report by the National Renewable Energy Laboratory, "Making Connections: Case Studies of Interconnection Barriers and Their Impact on Distributed Power Projects," which documents the results of a study of sixty-five distributed energy projects.[43] The various barriers—especially the countless obstacles and fees that utilities can place in the way of on-site generation projects—are well known to those of us who have been analyzing and advocating distributed energy projects for years. Anyone who believes that fuel cells will have an easy time entering the market for on-site power would do well to read the report. Its major findings were:

- Numerous technical, business practice, and regulatory barriers block distributed generation projects from being developed.
- Barriers include "lengthy approval processes, project-specific equipment requirements, and high standard fees."
- There is no national agreement on technical standards for grid interconnection, insurance requirements, and reasonable charges for interconnection or distributed generation.
- Vendors of distributed generation equipment can work to remove or reduce barriers, but this typically requires a great deal of time, effort, and money, which can render projects uneconomical.
- Distributed projects are typically not given "appropriate credit for the contributions they make to meeting power demand, reducing transmission losses, or improving environmental quality."

Certainly, fuel cells have some advantages that will minimize some of these barriers, particularly those associated with environmental permitting. Moreover, the federal government and some states, including California, are examining how they can reduce these barriers and give fuel cells and similar technologies credit for desirable attributes, such as reducing strain on the electric grid. But removing all these barriers will take many, many years. Indeed, Starwood itself faced utility efforts in 2003 to block deployment of a 250 kW molten carbonate fuel cell in a New York hotel; it succeeded in overcoming those efforts primarily because the system was relatively small, representing only 10 percent of the hotel's total power. In June 2003, Jerry Leitman, chief executive officer of Fuel-Cell Energy, described some of these barriers as "the ongoing battle between distributed generation and the local utility."[44]

Conclusion

More than 160 years since the technology was first demonstrated, fuel cells finally appear to be close to achieving the levels of cost and performance needed for success in the commercial marketplace. Nevertheless, the early markets for stationary fuel cells are smaller and less attractive than widely perceived, and the primary market is filled with multiple competitors and barriers to entry. Should high-temperature fuel cells, especially SOFCs, achieve the targets that are currently the focus of massive public and private R&D, they could by themselves usher in a fuel cell economy.

In this chapter, I have focused on the commercialization of stationary fuel cells because it may be relatively close at hand. Equally important, this is a market for which there is a great deal of real-world experience and data. In contrast, the path to commercialization for transportation fuel cells—and the ultimate realization of a hydrogen economy—will take much longer and is far more speculative. The chapters that follow will examine the technical and logistical challenges involved in bringing about a hydrogen economy.

CHAPTER 4

Hydrogen Production

Hydrogen is the best of fuels. Hydrogen is the worst of fuels.

On the plus side, hydrogen is the most abundant element in the universe by far. Hydrogen can be made from many things, which offers us the potential to replace our reliance on limited and insecure energy sources, such as oil, with the use of diversified, domestic, and, ultimately, limitless sources. Hydrogen can power highly efficient fuel cells that generate both electricity and heat with no emissions other than pure, drinkable water. Hydrogen represents one of the few substitutes for oil as a transportation fuel that will not contribute to global warming—if generated by renewable sources, such as wind power, or even by coal, should the capture and storage of carbon dioxide (CO_2) on a massive scale prove practical and affordable.

Hydrogen is widely used in industry today, and we have decades of experience in generating, storing, and transporting it. Although it has a reputation as a dangerous fuel, it has a good safety record in industrial settings.

On the minus side, hydrogen is not a readily accessible energy source, as are coal, oil, natural gas, sunlight, and wind.[1] Hydrogen is bound up tightly in such molecules as water and natural gas, so it is expensive and energy-intensive to extract and purify. Transportation fuel cell costs in 2003 exceeded internal combustion engine costs by far more than a factor of thirty. Hydrogen from renewable sources is especially expensive. The practicality of carbon sequestration is as yet unproven.

Hydrogen is difficult and costly to compress, store, and transport. It has one of the lowest energy densities of any fuel, one-third that of natural gas. It is expensive and energy-intensive to liquefy; moreover, a gallon of liquid hydrogen has only about one-quarter the energy of a gallon of gasoline. Hydrogen has major safety issues—it is flammable over a wide range of concentrations and has an ignition energy twenty times smaller than that of natural gas or gasoline—so leaks are a significant fire hazard. One of the most leak-prone of gases, hydrogen is subject to a set of strict and cumbersome codes and standards.

To paraphrase Charles Dickens again: We were all going directly to hydrogen, we were all going directly the other way. This chapter looks at options for producing hydrogen today and in the future— but first a little background on this most indispensable element.

A Brief History of Hydrogen

"If God did create the world with a word, the word would have been hydrogen," said Harlow Shapley, one of the twentieth century's greatest astronomers. In our current theory of the origin of the universe, when atoms first formed out of the sea of particles created in the big bang, basic hydrogen—one electron and one proton—constituted some 92 percent of the atoms, while virtually all of the rest was helium. Today, some 15 billion years later, about 90 percent of all particles are hydrogen and 9 percent are helium.[2]

Hydrogen is everywhere. In the near-vacuum of interstellar space, every cubic centimeter contains a few hydrogen atoms. In the interior of planet Jupiter, every cubic centimeter contains more than 10 million billion billion hydrogen atoms. Every second, our sun fuses 600 million tons of hydrogen into helium, providing the light and the warmth that make life on Earth possible. The hydrogen-driven process of stellar birth and death has generated all the heavier elements that make up our bodies and our planet, including the oxygen that, with hydrogen, composes water. As physicist John Rigden puts it in his delightful 2002 book *Hydrogen: The Essential Element,* "hydrogen is the mother of all atoms and molecules."[3]

Hydrogen's first "discovery" on Earth is often credited to Theophrastus Bombastus von Hohenheim (1493–1541), better known as Paracelsus, Renaissance physician and alchemist. Paracelsus observed that when acids react with metals, a flammable gas is emitted. But English nobleman Henry Cavendish (1731–1810) is typically given credit for discovering in 1766 that hydrogen is a separate substance and for characterizing a number of its qualities. He called hydrogen "inflammable air" and was the first chemist to produce water from oxygen and hydrogen. French scientist Antoine Lavoisier (1743–1794) learned of Cavendish's work in 1783 and repeated and expanded upon his experiments. Lavoisier, who died on the guillotine in 1794, named the element *hydrogène,* from the Greek words for "water" and "generate."

How remarkable, then, that hydrogen not only generates water when combined with oxygen but also helped generate the oxygen in stars in the first place. How doubly remarkable it would be if the sun's hydrogen-generated power were ultimately used to electrolyze that water to generate hydrogen to power the planet.

The idea of hydrogen as the ultimate fuel, limitless and powerful, stoked the imagination of nineteenth-century fiction writers. In his 1874 book *The Mysterious Island,* the ever-prescient writer Jules Verne has his characters discuss what would happen to America's "industrial and commercial movement" when the world runs out of coal in hundreds of years. The engineer, Cyrus Harding, explains that the world will turn to another fuel, "water decomposed into its primitive elements . . . doubtless by electricity, which will then have become a powerful and manageable force." Harding goes on to say:

> Yes, my friends, I believe that water will one day be employed as fuel, that hydrogen and oxygen which constitute it, used singly or together, will furnish an inexhaustible source of heat and light, of an intensity of which coal is not capable. Some day the coalrooms of steamers and the tenders of locomotives will, instead of coal, be

stored with these two condensed gases, which will burn in the furnaces with enormous calorific power. There is, therefore, nothing
to fear. As long as the earth is inhabited it will supply the wants of
its inhabitants, and there will be no want of either light or heat as
long as the productions of the vegetable, mineral or animal king-
doms do not fail us. I believe, then, that when the deposits of coal
are exhausted we shall heat and warm ourselves with water. *Water
will be the coal of the future.*[4]

Verne neglected to tell us where the primary energy to electrolyze the hydrogen would come from, but his prediction stands
as one of uncanny foresight.

A few years later, in 1893, British novelist Max Pemberton published a prophetic best seller, *The Iron Pirate*. In the book, a speedy
and powerful pirate ship terrorizes the Atlantic Ocean. The source
of its power baffles the narrator: "Neither steam nor smoke came
from her, no evidence, even the most trifling, of that terrible power
which was then driving her through the seas at such a fearful speed."
Ultimately, we learn that the ship is based in Greenland and hydrogen from coal fuels "the most powerful engines that have yet been
placed in a battle-ship." Interestingly, Pemberton writes, "the gas
itself was made by passing the steam from a comparatively small
boiler through a coke and anthracite furnace, the coke combining
with the oxygen and leaving pure hydrogen."[5] As we will see, generating hydrogen from coal is one of the principal areas of focus for
public and private sector research and development (R&D).

As the twentieth century dawned, hydrogen became an intense
focus of scientists, both theorists and experimentalists. "To understand hydrogen is to understand all of physics," one physicist said.
The study of hydrogen was critical to our understanding of the atom
and the evolution of the universe, to the development of quantum
mechanics and quantum electrodynamics, and to such practical
devices as magnetic resonance imaging (MRI) medical equipment.
Rigden's book is the best recent recounting of all this work.

In the middle of the twentieth century, scientists and governments expended tremendous effort in trying to replicate on Earth

the awesome power of the sun's hydrogen fusion. Strikingly, we have been all too successful in tapping fusion energy in uncontrolled reactions—unleashing the horrific power of the hydrogen bomb—but unsuccessful in efforts to create practical energy sources from a controlled fusion reaction to generate electricity. Controlled, earthbound fusion is still many decades away. For this century, the fusion energy we will most likely be tapping will be from the sun, in the form of its renewable radiating energy.

Efforts to harness hydrogen on a smaller scale, through direct combustion or fuel cells, gathered momentum throughout the twentieth century. An excellent history of those efforts can be found in Peter Hoffmann's 2001 book *Tomorrow's Energy*. The large investments made by the National Aeronautics and Space Administration to develop compact, lightweight engines and energy storage devices for space travel were especially important. As discussed in earlier chapters, making practical and affordable fuel cells has proven more difficult than expected. And even if we can develop transportation fuel cells that can potentially replace automobile engines, a true hydrogen economy will require the generation of enormous amounts of hydrogen from CO_2-free sources, which is not currently practical or affordable.

Hydrogen Generation: Today and Tomorrow

Hydrogen production is a large, modern industry with commercial roots reaching back more than a hundred years.[6] Globally, hydrogen is produced primarily for two purposes. The first is to synthesize ammonia (NH_3), especially for fertilizer production, by combining hydrogen with nitrogen. The second is hydro-formulation, or high-pressure hydro-treating, of petroleum in refineries, a process that, for instance, converts heavy crude oils into usable transportation fuels or that produces cleaner reformulated gasoline.

Global annual production is about 45 billion kilograms (kg) or 500 billion Normal cubic meters (Nm^3). A Normal cubic meter is a cubic meter at one atmosphere of pressure and 0°C. Hydrogen is produced through a variety of processes discussed below using traditional fuels (see Table 4.1).[7]

TABLE 4.1.
Global Hydrogen Production

Origin	Amount (billions of Nm³/year)	Percent
Natural gas	240	48
Oil	150	30
Coal	90	18
Electrolysis	20	4
TOTAL	500	100

Source: U.S. Department of Energy

Hydrogen production in the United States is currently about 8 billion kg (roughly 90 billion Nm³), or the energy equivalent of 8 billion gallons of gasoline. Hydrogen demand grew by more than 20 percent per year for much of the 1990s and has been growing at more than 10 percent per year since then, with the biggest demand growth coming from oil refineries.[8]

In 2000, Americans consumed nearly 180 billion gallons of gasoline, diesel fuel, and other transportation fuels for on-road travel. This represented about 20 percent of U.S. total energy consumption. More energy is consumed to power travel by air, water, and rail (and to power U.S. pipelines), bringing total transportation energy use alone to nearly 30 percent of all U.S. energy consumption.[9] The scale of hydrogen production in the United States and the world will have to be vastly greater than it now is just to replace transportation energy.

The rest of this chapter examines the current sources of hydrogen and possible future alternatives. Each option raises different questions involving its effects on hydrogen infrastructure options and on the energy and environmental problems we face. Many, if not most, of these questions cannot be answered today, but I will try to answer a number of them in subsequent chapters.

Natural Gas

Natural gas (methane, or CH_4) is by far the most common source of hydrogen.[10] Steam methane reforming (SMR) generates about half of global hydrogen and more than 90 percent of U.S. hydrogen (representing some 5 percent of U.S. gas consumption).[11] Conventional SMR is a multi-step process. In the first step, natural gas reacts with water vapor (H_2O) in tubes under high pressure (15 to 25 atmospheres) and high temperature (750°C–1,000°C), which are filled with a catalyst (typically nickel), forming carbon monoxide (CO) and hydrogen:

$$CH_4 + H_2O => CO + 3H_2$$

In the second reaction, called the water-gas shift, the CO is shifted with steam to produce CO_2 and extra hydrogen:

$$CO + H_2O => CO_2 + H_2$$

This reaction can be accomplished in one or two stages at temperature levels ranging from 200°C to 475°C.

The flue gas that leaves the shift reactor is some 70 to 80 percent hydrogen, together with CO_2, CH_4, water vapor, and CO. The hydrogen is then separated from the mixed gas stream for final use, typically using pressure swing adsorption (PSA), achieving high purity. (For proton exchange membrane, or PEM, fuel cells, a CO removal system is needed, since PEMs tolerate only a few parts per million of CO.) The CO_2 is usually vented to the atmosphere, but as concerns over global warming continue to grow, it might well be captured and sequestered. Both SMR and PSA are mature commercial processes. For SMR, the overall energy efficiency (the ratio of the energy in the hydrogen output to that in the fuel input) is about 70 percent.

Although SMR is by far the most widely used means of producing hydrogen on an industrial scale, small-scale SMR might also be used at local hydrogen filling stations. Currently, however, it is far too costly a process for providing a major source of hydrogen for transportation. Today, SMR plants have significant economies of scale—not only are large plants relatively cheaper to build (per unit output), but they also would very likely command a far lower price for natural gas than would smaller SMR plants at local urban filling stations. Partly offsetting that extra cost, local production would avoid the need to transport hydrogen from a central generation facility to the filling station (see Chapter 5). The cost of producing and delivering hydrogen from an SMR is currently projected to be $4 to $5 per kilogram, comparable to a gasoline price of $4–$5 per gallon.[12] Not surprisingly, a considerable amount of R&D is being focused on efforts to reduce the cost of SMR as well as its alternatives, such as partial oxidation and autothermal reforming.

While natural gas is both the least expensive source of hydrogen today and the most straightforward to scale up quickly for fueling PEM vehicles, we must answer these questions before pursuing this path:

- Is there enough natural gas to both meet the growing demand for gas-fired power plants and supply a significant fraction of a hydrogen-based transportation system?
- What would happen to the prices of natural gas, hydrogen, and electricity with a dramatic increase in the demand for natural gas to make hydrogen?
- Can the delivered cost of hydrogen from natural gas become competitive with the delivered cost of gasoline?
- Can the infrastructure costs be reduced to manageable levels? Current estimates are as much as a trillion dollars or more.
- Which will be cheaper and/or more practical, reforming methane at small local filling stations or at large centralized plants? Could technological advances change the answer to that question?
- Are the global warming benefits from methane-based hydrogen sufficient to justify building an infrastructure around SMRs, or should we wait until we can build the infrastructure around a CO_2-free source of hydrogen?

These basic questions are in addition to the fundamental question that applies to all means of generating hydrogen: Can automakers build an affordable and practical PEM vehicle that will use the hydrogen?

Water

Water is another common source of hydrogen. Electrolysis, the process of decomposing water into hydrogen and oxygen using electricity, is a mature technology widely used around the world to generate very pure hydrogen. It is, however, an extremely energy-intensive process, and the faster you want to generate hydrogen, the more power you need per kilogram produced. Typical commercial electrolysis units require about 50 kWh per kilogram, which represents an energy efficiency of 70 percent—that is, more than 1.4 units of energy must be provided to generate 1 energy unit in the hydrogen. And since most electricity comes from fossil fuels, and the average fossil fuel plant is about 30 percent efficient, the overall system efficiency is close to 20 percent (70 percent times 30 percent)—four units of energy are thrown away for every one unit of hydrogen energy produced. That is a lot of energy to waste.

As with SMR plants, larger electrolysis plants are relatively cheaper to build (per unit output)—and they would very likely command a far lower price for electricity—than smaller ones at local urban filling stations (sometimes called "forecourt plants" because they are based right where the hydrogen is needed). Hydrogen could be generated at low nighttime off-peak rates, but that is easier to do at a centralized production facility than at a local filling station, which must be responsive to customers who typically do most of their fueling during the day and early evening, peak power demand times. "To circumvent peak power rates," Dale Simbeck and Elaine Chang noted in their July 2002 analysis for the National Renewable Energy Laboratory (NREL), "forecourt plants have to be built with oversized units operated at low utilization rates with large amounts of storage. This option would require considerable additional capital investment."[13]

Simbeck and Chang estimate the cost today of producing and delivering hydrogen from a central electrolysis plant at $7–$9/kg. They put the cost of production at a forecourt plant at $12/kg. High cost is probably the main reason why only a small percentage of the world's current hydrogen production comes from electrolysis. Moreover, *to replace all the gasoline sold in the United States today with hydrogen from electrolysis would require more electricity than is sold in the United States today.*[14]

From the perspective of global warming, electrolysis makes little sense for the foreseeable future because both electrolysis and central-station power generation are relatively inefficient processes, and most U.S. electricity is generated by the burning of fossil fuels. Burning a gallon of gasoline releases about 20 pounds of CO_2. Producing 1 kg of hydrogen by electrolysis would generate, on average, 70 pounds of CO_2.[15] A gallon of gasoline and a kilogram of hydrogen have about the same energy, and even allowing for the potential doubled efficiency of fuel cell vehicles, producing hydrogen from electrolysis will make global warming worse. Because of these economic and environmental problems, it seems unlikely that the nation will pursue generation of significant quantities of hydrogen from the U.S. electric grid anytime soon.

Hydrogen could be generated from renewable electricity, but the renewable system most suitable for local generation—solar photovoltaics—currently makes hydrogen that is far too expensive. The least expensive form of renewable energy—wind power—still constitutes only a few tenths of 1 percent of all U.S. generation, although that figure is rising rapidly.[16] These basic questions need to be answered before we pursue this enticing path:

- Is generating hydrogen from electrolysis powered by renewables a good use of that power from economic and environmental perspectives?
- How long will it be before the United States has enough excess low-cost renewable generation that it can divert a substantial fraction to production of hydrogen?
- What are the prospects that forecourt hydrogen generation from solar photovoltaics will be wise or practical in the first half of the century?
- If hydrogen is generated from the vast wind resources in the Midwest, what would be the infrastructure costs for delivering it?

Gasoline

Gasoline can be used as a source of hydrogen. Hydrogen can be produced from hydrocarbons such as gasoline (and methane) with partial oxidation and autothermal reformers.[17] The toughest practical problem for onboard gasoline reformers—other than high cost—is that the high temperature at which they operate does not allow a rapid start for the automobile, a design feature we have all come to expect.

In May 2003, Nuvera Fuel Cells announced that in conjunction with the U.S. Department of Energy (DOE) and a European automaker, the company had demonstrated a 75 kW gasoline reformer with more than 80 percent efficiency.[18] This device cannot start up in less than a minute, but the company is working on a next-generation device aimed at beating a thirty-second start time. Since the 75 kW reformer is not a commercial product, the company did not announce its price. Prashant Chintawar, Nuvera's executive director for business development and strategic R&D, has told me that it would be at least ten years before the device would achieve a cost of $25–$30/kW, the range needed for an affordable car.

When I was at the DOE in the mid-1990s, we thought this area of R&D was valuable because it was far from clear that we could solve all the technological problems related to building a pure hydrogen infrastructure, including practical and affordable onboard storage of hydrogen, or that we could solve the chicken-and-egg problem between the hydrogen fueling infrastructure and the hydrogen-powered vehicles. Gasoline fuel cell vehicles (FCVs) seemed like a worthwhile interim step. As a comprehensive 2001 study on commercializing fuel cell vehicles concluded, "a major potential advantage of gasoline FCVs is the prospect of a conventional fuel . . . that requires essentially no infrastructure changes or investment for FCV use." The study, undertaken for the California Fuel Cell Partnership (CaFCP), notes, "It would also eliminate the new health and safety concerns of other fuels."[19]

Nonetheless, many people questioned this strategy when it was first developed, and many continue to do so. For instance, following the 2003 Nuvera announcement, David Redstone, editor and publisher of the *Hydrogen & Fuel Cell Investor's Newsletter*, wrote:

> What is the point of a gasoline reformer? Isn't the point of "the hydrogen economy" to eliminate carbon from the fuel chain (thereby eliminating CO_2 emissions) and to end our dependence on foreign energy sources (Saudi-Kuwaiti-Iraqi oil)? How is switching from burning gasoline in ICEs [internal combustion engines] to reforming it for use as H_2 in fuel cells going to get us where we need to go, especially when the fuel cells currently being contemplated for car engines are barely more efficient than ICEs and the reforming process involves a loss of 20 percent of the energy contained in the gasoline?[20]

These are reasonable questions. Several additional questions need to be answered before we pursue a path of generating large volumes of hydrogen from gasoline reformers on board fuel cell vehicles:

- Can we build an affordable and practical gasoline reformer?
- Is the chicken-and-egg problem solvable, or will we need this technology to achieve significant market penetration of fuel cell vehicles before we build the hydrogen infrastructure?
- Are the net energy and environmental benefits of gasoline-powered fuel cell vehicles worth the extra cost of the system?

Methanol

Methanol, also known as wood alcohol, is another widely discussed potential source (or carrier) of hydrogen.[21] Chemically, methanol, CH_3OH, is a clear liquid, the simplest of the alcohols, with one carbon atom per molecule. Methanol is extensively used today—U.S. demand in 2002 exceeded 2 billion gallons. The largest U.S. methanol markets are for producing the gasoline additive MTBE (methyl tertiary butyl ether) as well as formaldehyde and acetic acid.

Methanol is already used as a transportation fuel. It has been the fuel of choice at the Indianapolis 500 for more than three decades, in part because it improves the performance of the cars but primarily because it is considered much safer. It is less flammable than gasoline and, when it does ignite, causes less severe fires; one 1990 study for the U.S. Environmental Protection Agency (EPA) concluded, "Pure methanol is projected to result in as much as a 90 percent reduction in the number of automotive fuel related fires relative to gasoline." Methanol appears to biodegrade quickly when spilled. It dissolves and dilutes rapidly in water. It has been actively promoted as an alternative fuel by the EPA and the DOE, in part because it has reduced urban air pollutant emissions more effectively than gasoline. Most methanol-fueled vehicles use a blend of 85 percent methanol and 15 percent gasoline called M85.

Methanol has a number of advantages for powering fuel cell vehicles. As the 2001 study for the CAFCP noted, these include methanol's "immediate availability without new upstream infrastructure, high hydrogen-carrying capacity, and ability to be readily stored, delivered, and carried on-board without pressurization."[22] In short, our transportation system and its infrastructure favor liquid fuels. Fuel cell vehicles with onboard methanol reformers would have very low emissions of urban air pollutants. Daimler-Chrysler has introduced demonstration fuel cell vehicles that convert methanol to hydrogen on board.

A key advantage of methanol is versatility. Methanol reformers operate at much lower temperatures ($250°C–350°C$), so they are more practical than onboard gasoline reformers. Also, methanol reformers could be used at fueling stations to generate forecourt hydrogen. Most intriguingly, considerable research is aimed at developing a direct methanol fuel cell (DMFC), which could run on methanol without a reformer—although practical, affordable DMFCs for cars and trucks appear a long way off. And while methanol is primarily synthesized from natural gas, it can also be produced from a number of CO_2-free sources, including municipal solid waste and plant matter.[23]

On the other hand, many features of methanol would make its widespread use as a transportation fuel problematic. It is very toxic. As the EPA has noted, "a few teaspoons of methanol consumed orally can cause blindness, and a few tablespoons can be fatal, if not treated." Methanol is also very corrosive, so it requires a special fuel-handling system. Most major oil companies currently responsible for delivering our transportation fuels are at best unenthusiastic about creating a methanol economy, and some have expressed outright opposition.

Some environmentalists and former environmental regulators I have spoken to are reluctant to embrace a dramatic increase in methanol use, in part because it is used to make MTBE, a gasoline additive now being phased out in California because of environmental concerns such as groundwater contamination (although in fairness to methanol, which exists in nature and degrades quickly, MTBE, in contrast, is a complex, man-made compound that exhibits little degradation once released into the environment).

For any dramatic increase in U.S. methanol consumption, most of the supply would have to be imported. While biomass-generated methanol might be economical in the long term, there is a considerable amount of so-called stranded natural gas in distant locations around the globe that could be converted to methanol and shipped by tanker at relatively low cost, should increased demand warrant such investment.[24] Methanol from natural gas would have little or no net greenhouse gas benefits in a future fuel cell vehicle, as compared with future hybrid electric vehicles (see Chapter 8).

These questions need answering before we pursue a path of building a transportation system around a methanol economy:

- Is there enough natural gas both to meet the growing demand for gas-fired power plants and to supply a significant fraction of a transportation system built around methanol?
- Can the price of methanol remain competitive with that of gasoline if methanol demand increases sharply in the coming decades?
- What would be the overall health and safety effect of large-scale use of methanol as a consumer product?
- What are the prospects for generating a substantial fraction of methanol cost-effectively from renewable or CO_2-free sources?
- Will direct methanol fuel cells be affordable and practical anytime soon?

- Can methanol achieve the necessary support from businesses and consumers to be viable as the nation's primary transportation fuel?

Coal

Coal is a major source of hydrogen. To produce hydrogen, typically coal is gasified, impurities are removed, and then its hydrogen is recovered. This process results in significant emissions of CO_2, and so, from a global warming perspective, current coal gasification technology could not serve as a basis for a hydrogen economy. In addition, Simbeck and Chang estimate that producing and delivering coal-generated hydrogen would cost \$4.50–\$5.60/kg, several times the cost of U.S. gasoline on an equivalent energy basis.

Coal is the most abundant fossil fuel in the United States and a great many other countries. For instance, coal constitutes some 95 percent of the nation's fossil energy reserves, and it is estimated that, with the use of existing technology, known recoverable coal reserves could sustain the nation's current level of coal consumption for more than 300 years.[25] Major developing countries such as China and India also have vast coal reserves. However, burning any significant fraction of global coal reserves using current power plant technology would almost certainly cause devastating and irreparable harm to the global climate (see Chapter 7). Hence, many countries and companies are pursuing R&D into generating hydrogen, electricity, or both from coal without releasing CO_2.

Here is one strategy being pursued by the DOE: Use a gasification and cleaning process that combines coal, oxygen (or air), and steam under high temperature and pressure to generate a synthesis gas (syngas) made up primarily of hydrogen and CO, without impurities such as sulfur or mercury.[26] The water-gas shift reaction described earlier is then applied to increase hydrogen production and create a stream of CO_2 that can be removed and piped to a sequestration site (see Chapter 8). The rest of the hydrogen-rich gas is sent to a PSA system for purification and transport. The remaining gas that comes out of the PSA system can be compressed and sent to a combined cycle power plant (similar to the natural gas combined cycle plants now in widespread use). Hydrogen can also power a combined cycle plant (as can syngas), so this system can be configured to generate more hydrogen and less electricity or vice versa.

In February 2003, the Department of Energy announced FutureGen (the Integrated Sequestration and Hydrogen Research Initiative), a ten-year, billion-dollar project to design, build, and construct a 275 MW prototype plant that would cogenerate electricity and hydrogen and sequester 90 percent of the CO_2. The goal of the system is to "validate the engineering, economic, and environmental viability of advanced coal-based, near-zero emission technologies that by 2020" will produce electricity that is only 10 percent more expensive than current coal-generated electricity and produce hydrogen that is competitive in price with gasoline (although this does not include the cost of delivering the hydrogen). Initially, the hydrogen would be used to generate clean power, perhaps using a solid oxide fuel cell (SOFC).[27]

A 2002 study for the National Energy Technology Laboratory found that coal gasification systems with CO_2 capture could achieve efficiencies of 60 percent or more in cogenerating hydrogen and electricity using various combinations of turbines and SOFCs.[28] Nonetheless, just as commercializing fuel cells has taken much longer and has proven far more difficult than was expected, so, too, may building large commercial coal gasification combined cycle units. One 2003 analysis described the "embarrassingly poor history" traditional power generators have had with far simpler chemical processes.[29]

Sequestering the CO_2 will be another technological challenge. Sequestration is the process of locking up CO_2 (for instance, in large underground formations) so it cannot enter Earth's atmosphere. While CO_2 separation and capture are a common part of many industrial processes, existing technologies would not be cost-effective for large-scale sequestration. Estimates of sequestration costs using current technology are exceedingly high. Moreover, the practicality and environmental consequences of many pathways for sequestration are not yet proven from an engineering or scientific perspective. Sequestration remains a focus of active research and development and is discussed in greater detail in Chapter 8.

These questions need answering before we can pursue the path of generating large volumes of hydrogen from coal and sequestering the CO_2 produced:

- Can cogeneration of hydrogen and electricity from coal, coupled with CO_2 extraction, be made into an affordable and practical system for cost-effectively generating both energy carriers?
- If hydrogen is generated from large coal plants outside cities, perhaps close to existing coal mines, what would be the infrastructure costs for delivering the hydrogen to consumers?
- Can CO_2 sequestration on a massive scale be practical, economical, and permanent?

Biomass

Biomass could become a major source of hydrogen.[30] Biomass is any material that has participated in the growing cycle. It includes agricultural, food, and wood waste as well as trees and grasses grown as energy crops. Biomass is of interest from a global warming perspective because when it is transformed into energy, by burning, for instance, it releases CO_2 that was previously sequestered from the atmosphere (temporarily) in the growing cycle, so the net CO_2 emitted is zero.[31] Biomass thus represents a potentially sustainable way of providing energy without accelerating global warming.

Biomass can also be converted to a liquid fuel, such as ethanol, which is then used as a gasoline blend. Today, the major biofuel is ethanol produced from corn, which yields only about 25 percent more energy than was consumed to grow the corn and make the ethanol.[32] The future holds the promise of ethanol from sources other than corn, dedicated energy crops such as switchgrass, which can be grown and harvested with minimal energy consumption so that overall emissions are near zero (see Chapter 8).

Biomass is of special interest because it seems likely to be the lowest-cost renewable source of hydrogen in the near term and because it may be a viable renewable source of hydrogen on a large scale in the long term. Biomass can be gasified and converted into hydrogen (and electricity) in a process very similar to coal gasification, described earlier. A number of biomass gasification processes are being demonstrated, although a practical commercial system remains technologically challenging. Biomass can also be gasified together with coal; the Royal Dutch/Shell Group has commercially demonstrated a 25/75 biomass/coal gasifier.

Perhaps most intriguing, if CO_2 sequestration is possible, one could imagine extracting the CO_2 stream from the biomass gasification process, as is being contemplated for coal gasification. This would turn biomass into a potent net reducer of CO_2—it would mean extracting CO_2 from the air while growing and then injecting that CO_2 into underground reservoirs through the gasification and sequestration process.

Another promising approach is pyrolysis, the use of heat to decompose biomass into its constituents. The ultimate idea is to create a "bio-refinery," analogous to a petroleum refinery, where biomass is converted into many different useful products. In one version of this process, biomass is dried and heated, coproducts are removed, and the hydrogen is produced via steam reforming.[33]

Like coal, biomass seems unlikely to serve as a source of small-scale on-site hydrogen production. Simbeck and Chang estimate the cost of delivered hydrogen from biomass gasification at $5.00–$6.30/kg, depending primarily on the means of delivery. Other studies by NREL suggest a lower cost, especially for pyrolysis, should we achieve significant technological improvements and successful commercialization of biomass and hydrogen infrastructure technologies.[34]

Waste biomass, such as peanut shells or bagasse (the residue from sugarcane), tends to be the most cost-effective source, but the ultimate supply is limited. Even in a country with as much arable land as the United States, a large fraction of agricultural land would need to be devoted to biomass production if that were to serve as the major source of transportation fuel.

These questions need answering before we can consider biomass as a likely source for generating large volumes of hydrogen renewably:

- What fraction of arable land in the United States (and the world) would be needed for biomass-to-hydrogen production sufficient to displace a significant fraction of gasoline—and is that a practical or politically feasible approach?
- Can hydrogen from biomass be made cost-competitive with gasoline and with other sources of hydrogen?
- If hydrogen is generated from large biomass plants far away from cities, what would be the infrastructure costs for delivering the hydrogen to consumers?
- Is hydrogen generation the best use for biomass from an environmental or an energy security perspective?

Nuclear Power Plants

Nuclear power plants can also be a source of hydrogen.[35] Straight electrolysis of water with electricity from a nuclear power plant is one possibility, but that seems unlikely to be economically efficient for the foreseeable future. On the other hand, nuclear power plants, like most central-station power plants, generate a considerable amount of heat that they end up throwing away. This creates an opportunity because it is more efficient to generate hydrogen at higher temperatures, for instance by thermochemically decomposing water into hydrogen and oxygen through a set of chemical reactions.

Thermochemical water-splitting processes at temperatures exceeding 750°C could theoretically achieve 40 to 52 percent efficiency in hydrogen production.[36] Cogeneration of electricity could raise the overall efficiency to as high as 60 percent. The fact that a process could be efficient, however, does not mean it would be economical, and we are a long way from knowing whether this approach can compete with other emerging hydrogen generation technologies. The DOE is currently pursuing thermochemical hydrogen production systems using nuclear power, with the goal of demonstrating commercial-scale production by 2015.

No plausible strategy for generating hydrogen that is potentially CO_2-free can be dismissed out of hand, but I am skeptical that nuclear-generated hydrogen could be a practical solution, at least in the United States, especially since 100 or more nuclear water-splitting plants would be needed to replace a significant fraction of U.S. transportation fuel with hydrogen. A major 2003 interdisciplinary study by the Massachusetts Institute of Technology, *The Future of Nuclear Power*, highlighted many of the "unresolved problems" that have created "limited prospects for nuclear power today." The study found that "in deregulated markets, nuclear power is not now cost competitive with coal and natural gas." The public has significant concerns about safety, environmental, health, and terrorism risks associated with nuclear power. The study also found that "nuclear power has unresolved challenges in long-term management of radioactive wastes." The study described possible technological and other strategies for addressing these issues, but noted, for instance that "the cost improvements we project are plausible but unproven."[37]

These questions need answering before we pursue a path of generating large volumes of hydrogen from nuclear power:

- Can nuclear power be a safe and economical source of hydrogen capable of attracting enough investment capital to build so many new plants?
- If hydrogen is generated from nuclear power plants outside cities, what would be the infrastructure costs for delivering the hydrogen to consumers?
- How much additional nuclear power capacity would need to be added to meet the needs of the hydrogen economy? How would the nuclear waste be disposed of?

Other Sources of Hydrogen

Novel strategies for generating hydrogen abound. For instance, research is being conducted into using solar energy to directly facilitate the production of hydrogen, such as by using photovoltaic devices to split water directly. Considerable research is also taking place into producing hydrogen through active biological processes, such as adapting photosynthesis processes for hydrogen production and using existing or bioengineered bacteria to decompose organic compounds into hydrogen. As noted in an editorial in the special 2002 issue of the *International Journal of Hydrogen Energy* devoted to biohydrogen, this is still a relatively small and long-term area of research because "so far efficiencies obtained are low, productivity is low and costs are high compared to alternative technologies."[38]

Finally, fuel cells themselves, especially high-temperature ones, could trigenerate electricity, heat, and hydrogen. FuelCell Energy Inc. is pursuing this strategy for its molten carbonate fuel cell, as are solid oxide fuel cell (SOFC) companies. Fuel cells running on natural gas typically use about three of the four hydrogen atoms in methane (CH_4) for power generation. The remaining hydrogen goes into the flue gas or stack effluent with various amounts of CO_2, CO, and water vapor, depending on the type of fuel cell. That flue gas is sometimes vented to the atmosphere, sometimes combusted for heat, and sometimes used to facilitate the reforming process.

Hydrogen could, however, be separated and purified from the flue gas at a relatively low cost if high-temperature fuel cells can be made into successful commercial products. That is, when an SOFC or molten carbonate fuel cell can cost-effectively generate electricity and heat, the fuel (natural gas) and the fuel cell are already paid for. The cost of the hydrogen generation would then be just the incremental cost of the purification and separation system. A future SOFC system might cogenerate hydrogen for about $2.00 per kg.[39] This is still somewhat more expensive than gasoline, but it is potentially a very attractive price, since these fuel cells could be sited at an urban fueling station, thus avoiding the need for a costly hydrogen delivery infrastructure.

This promising strategy will, however, require fuel cells to have achieved significant technological and cost improvements while overcoming the numerous barriers to commercialization discussed in Chapter 3.

Conclusion

There are many different promising technologies and strategies for producing hydrogen, each with its own great advantages and severe disadvantages. It is possible that one will come to dominate. It is possible that multiple means will be used. These are still very early days in the transition to a hydrogen economy, and no one can know.

The most cost-effective existing technologies are, unfortunately, the ones that generate the most greenhouse gas pollution and thus face the greatest risk of stranded investment—of needing to be replaced when government policies change, as I believe they inevitably will (see Chapters 7 and 8). For instance, if by 2020 we were to build a hydrogen infrastructure around small steam methane reformers in local fueling stations, and then decide that greenhouse gas emissions must be dramatically reduced, we would have to replace that infrastructure almost entirely.

At this time, no set of commercial technologies appears able to deliver hydrogen to vehicles at a price much below about four times the current cost of gasoline (untaxed) on an equivalent energy basis. Nor does onboard reforming of gasoline appear likely to be a cost-effective and practical interim strategy without significant technological advances. Centralized hydrogen production far away from cities is the least expensive way of generating hydrogen for the foreseeable future, but that would entail a considerable investment in a delivery infrastructure, which is the subject of the next chapter.

CHAPTER 5

Key Elements of a Hydrogen-Based
Transportation System

Bernard Bulkin, chief scientist for British Petroleum, discussed BP's experience with its customers at the National Hydrogen Association's annual conference in March 2003. He said, "One thing that we know is that if hydrogen is going to make it in the mass market as a transport fuel, it has to be available in 30 to 50% of the retail network from the day the first mass manufactured cars hit the showrooms." He went on to say, "We know that the customer must be able to fill the tank in about a minute. Safely, with no leaks, while telling the kids to keep quiet in the back of the car."[1]

That would mean 50,000 to 90,000 newly fitted service stations. And, unless hydrogen is generated on-site at those fueling stations, we will also need a massive infrastructure for delivering that hydrogen from wherever it is generated. To accomplish such a seismic change, we will need major advances in hydrogen production and delivery, since *a hydrogen fueling infrastructure alone based on current commercial or near-commercial technology could cost more than a half trillion dollars.*[2]

Instead of filling the tanks of our cars and trucks with gasoline, a liquid with a very high energy density (delivered energy per unit volume), we would fill up with hydrogen, a gas with a very low energy density. Our cars will need not merely a wholly new engine, most likely a fuel cell, but also a wholly new onboard storage tank capable of carrying enough hydrogen to satisfy our demand for vehicles with a good driving range from one tank of fuel. No current commercial technology satisfies the demands of cost, volume, and weight needed for onboard hydrogen storage.

Finally, while hydrogen is safer than gasoline in many ways—it doesn't splatter or pool—it poses its own unique set of safety hazards, which will require new codes, new standards, and new technology. This chapter examines the key elements of the transition to a hydrogen economy. I will also begin answering some of the questions posed in the previous chapter, trying to sort out plausible scenarios from less plausible ones. First, however, some recent history.

Alternative Fuel Vehicles in the 1990s

Achieving a pure hydrogen economy will almost certainly require major government regulations and incentives, particularly to address the chicken-and-egg problem: Which will come first, hydrogen vehicles or the hydrogen fueling delivery system? As one 2003 article titled "Initiating Hydrogen Infrastructures" noted, "it is highly unlikely that market forces alone could result in the rapid and coordinated installation of thousands of [hydrogen fueling] stations spread uniformly across interstates and metropolitan areas."[3] Let us briefly examine lessons learned from efforts by the federal government to promote alternative fuel vehicles in the 1990s.

In 1992, the United States Congress passed the Energy Policy Act, and President George H. W. Bush signed it into law.[4] One of the goals was to reduce the amount of petroleum used for transportation by promoting the use of alternative fuels in cars and light trucks. These fuels include natural gas, methanol, ethanol, propane, electricity, and biodiesel. Alternative fuel vehicles (AFVs) operate on these fuels, although many are dual-fueled; that is, they can also run on gasoline.

The act established two goals in this area—having alternative fuels replace at least 10 percent of petroleum fuels in 2000 and at least 30 percent in 2010. The act mandated that a portion of new vehicles purchased for state and federal government fleets, as well as alternative fuel providers, must be AFVs. The act also gave the U.S. Department of Energy (DOE) the authority to promote AFVs in a variety of other ways, such as by developing voluntary partnerships with city governments and others. The job of managing this work fell to the Office of Energy Efficiency and Renewable Energy, the office I helped oversee and then managed during much of the 1990s. The initial results of this effort were documented in a February 2000 report by the General Accounting Office (GAO).

By 1999, some 1 million AFVs were on the road, only about 0.4 percent of all vehicles. In 1998, alternative fuels consumed in AFVs substituted for some 334 million gallons of gasoline, about 0.3 percent of that year's total consumption. An additional 3.9 billion gallons of ethanol and methanol replaced gasoline that year, but those fuels were blended with gasoline and consumed in conventional vehicles.

These results are far from a resounding success. My old office at the DOE is exceedingly good at developing clean energy technologies and getting people to use more efficient versions of existing technology (more efficient lighting, motors, heating and cooling). Our efforts have been verified by the National Academy of Sciences as having saved businesses and consumers more than $30 billion in energy costs. But getting people to use alternative fuel vehicles proved to be an exceedingly difficult task. The GAO, which is often critical of the work of government agencies, did not criticize the DOE's actions; instead, it concluded:

> The goals in the act for fuel replacement are not being met principally because alternative fuel vehicles have significant economic disadvantages compared to conventional gasoline vehicles. Fundamental economic impediments—such as *the relatively low price of gasoline, the lack of refueling stations for alternative fuels, and the additional cost to purchase these vehicles*—explain much of why both mandated fleets and the general public are disinclined to acquire alternative fuel vehicles and use alternative fuels.

A 2002 analysis of the efforts to commercialize natural gas vehicles (NGVs) in Canada and the United States had additional conclusions. The environmental benefits of NGVs were oversold, as were early cost estimates for both the vehicles and the fueling stations: "Early promoters often believe that 'prices just have to drop' and cited what turned out to be unachievable price levels." The study concluded, "Exaggerated claims have damaged the credibility of alternate transportation fuels, and have retarded acceptance, especially by large commercial purchasers."[5]

With these sober lessons in mind, let's examine some of the technology and infrastructure issues related to advancing hydrogen as an alternative fuel.

Infrastructure and Storage

A pure hydrogen economy will require a sea change in the fueling infrastructure that has built up over the past century to service the internal combustion engine. Two key issues will determine much of the nature of that infrastructure: where the hydrogen is produced and in what form it is stored on board the hydrogen vehicle. In the previous chapter, we saw that there are three likely places where hydrogen could be produced.

First, hydrogen could be generated on the car or truck itself, most likely by a gasoline or methanol reformer. Gasoline and methanol are very potent ways of storing hydrogen—both contain more hydrogen per gallon than even liquid hydrogen. If onboard reforming proves to be practical, the rest of the infrastructure will be much easier, since it will require only delivering a liquid fuel to a fueling station and then filling up a car with the liquid fuel, exactly as we do today. If small gasoline reformers were viable, we would need essentially no new infrastructure, although, as has been discussed, the environmental and energy security benefits of this strategy are modest and the technical challenges imposing, and therefore very few automakers and proton exchange membrane (PEM) fuel cell companies are pursuing this option. I will discuss some of the issues surrounding a methanol infrastructure later in this chapter.

Second, hydrogen could be produced at fueling stations located in cities and on highways. This is sometimes called forecourt production. Producing hydrogen in your home, however, seems impractical, as discussed in the next chapter.

Third, hydrogen could be produced at centralized facilities nearer to potential fuel sources, such as coal plants or windmills. In order for a car to be run on pure hydrogen, however, it must be able to safely, compactly, and cost-effectively store hydrogen on board, which is a major technical challenge. Under most conditions, hydrogen has a far lower energy-to-volume ratio than do fuels such as methane, methanol, propane, and octane (the principal component of gasoline). That is, hydrogen contains much less energy per gallon than other fuels at the same pressure. At room temperature and pressure, hydrogen takes up approximately 3,000 times more space than gasoline containing an equivalent amount of energy. Hydrogen does have an exceptional energy content per unit mass, such as per kilogram (kg), nearly triple that of gasoline, but the storage equipment on a car fitted for hydrogen use, such as pressurized tanks, adds significant weight to the system and negates this advantage.

The DOE's 2003 *Fuel Cell Report to Congress* notes:

> Hydrogen storage systems need to enable a vehicle to travel 300 to 400 miles and fit in an envelope that does not compromise either passenger space or storage space. Current energy storage technologies are insufficient to gain market acceptance because they do not meet these criteria.[6]

The driving range requirement will probably require a tank holding about 5 kg of hydrogen or more, depending on the size and weight of the vehicle. At the same time, we expect a vehicle that can be fueled in well under five minutes, with a storage system that is safe, that doesn't leak, and, of course, that is lightweight and affordable. Quite a design challenge.

The primary categories of hydrogen storage are physical and chemical. Physical storage includes liquefied hydrogen and compressed hydrogen gas. Chemical storage includes the formation of metal hydrides. These are seen to be the most likely near-term options.

Liquid Hydrogen

Liquid hydrogen is widely used today for storing and transporting hydrogen. Liquids enjoy considerable advantages over gases from a storage and fueling perspective. Compared with gases, they have a high energy density, are easier to transport, and are typically easier to handle. Hydrogen, however, is anything but typical. At atmospheric pressure, hydrogen becomes a liquid only at the ultra-frigid temperature of $-253°C$ ($-423°F$), just a few degrees above absolute zero. It can be stored only in a super-insulated tank, known as cryogenic storage.

The National Aeronautics and Space Administration (NASA) uses liquid hydrogen, along with liquid oxygen, as a fuel for the space shuttle. Some 100 tons, or nearly 400,000 gallons, of liquid hydrogen are stored in the shuttle's giant external tank.[7] To fuel each shuttle launch, fifty tanker trucks drive several hundred miles from New Orleans to Kennedy Space Center in Florida. We have a great deal of experience in shipping liquid hydrogen: Since 1965, NASA has trucked more than 100,000 tons of liquid hydrogen to Kennedy and Cape Canaveral. Another advantage of liquid hydrogen is that it can be stored in vessels that are relatively compact and lightweight. The General Motors Corporation (GM) has described a 90 kg cryogenic tank holding 4.6 kg of liquid hydrogen in a total volume of 34 gallons.[8] It may be possible to build a lighter tank holding more liquid hydrogen.[9]

With all these virtues, however, liquid hydrogen has a number of major disadvantages. The process of liquefying hydrogen requires expensive equipment and is very energy-intensive. Refrigeration processes have inherent efficiency limitations, and hydrogen liquefaction requires multiple stages of compression and cooling. Some 40 percent of the energy of the hydrogen is required to liquefy it for storage.[10] Moreover, the smaller the liquefaction plant, the more energy-intensive the system required, making it even less practical for local fueling stations.[11]

A major challenge associated with liquefied hydrogen is evaporation. Hydrogen stored as a liquid can boil off and escape from the tank over time. NASA faces this in the extreme: The agency loses almost 100,000 pounds of hydrogen each time it fuels up the shuttle, requiring it to truck in far more hydrogen than the 227,000 pounds needed by the main tank. In an automobile, this effect is particularly severe when it remains idle for a few days. The GM tank has a boil-off rate of up to 4 percent per day. Automakers are pursuing strategies for bleeding off the evaporating hydrogen, but this adds system complexity and will undermine safety and cost-effectiveness if that hydrogen cannot be utilized productively.

The cost of cryogenic storage—including the cost of the storage tank, the cost of the equipment to liquefy hydrogen, and the cost of electricity to run the equipment—remains high. Liquid hydrogen requires extreme precautions in handling because it is at such a low temperature. Fueling is typically done mechanically, such as with a robot arm.

From a global warming perspective, even with large, centralized liquefaction units, the electricity consumed would be quite high. According to Raymond Drnevich of Praxair Inc., a leading supplier of liquefied hydrogen in North America, the typical power consumption is 12.5 to 15 kilowatt-hours (kWh) per kilogram of hydrogen liquefied.[12] Since that electricity would come from the U.S. electric grid, liquefying 1 kg of hydrogen would by itself release some 17.5 to 21 pounds of carbon dioxide (CO_2) into the atmosphere for the foreseeable future. Burning 1 gallon of gasoline, which has roughly the same energy content as 1 kg of hydrogen, releases about the same amount—20 pounds—of CO_2 into the atmosphere. So even allowing for the greater efficiency of hydrogen fuel cell vehicles, if liquefaction is a major part of the hydrogen infrastructure, it would be exceedingly difficult for hydrogen-fueled vehicles to have a net greenhouse gas benefit until the electric grid were far "greener" than today (that is, has far lower CO_2 emissions per kilowatt-hour).

Compressed Hydrogen

Compressed hydrogen has been used in demonstration vehicles for many years, and nearly all prototype hydrogen vehicles today use this means of storage. Hydrogen compression is a straightforward process using a mature technology. It is relatively low in cost compared with liquefaction and emerging technologies.

Hydrogen is compressed up to pressures of 3,600 pounds per square inch (psi), 5,000 psi, or even 10,000 psi. For comparison, atmospheric pressure is about 15 psi. Even at these high pressures, hydrogen has a much lower energy per unit volume than gasoline does. Increasing compression allows more fuel to be contained in a given volume and therefore increases the energy density. However, it also requires a greater energy input. Compression to 5,000 or 10,000 psi is a multi-stage process that could require energy input equal to 10 to 15 percent of the fuel's usable energy content.[13] Research focuses on developing systems that are strong and lightweight, able to contain hydrogen under extreme pressures (even in the event of a collision) without adding excessive weight to the vehicle.

Compressed hydrogen can be fueled relatively fast, and the tanks can be reused through many cycles without any deterioration of capacity or performance. The main technical challenges associated with compressed hydrogen are the weight of the storage tank and the volume required (safety issues are addressed later in this chapter). Tank weight will improve with the development of strong and lightweight containment vessels. Tank volume is addressed by using the highest-pressure systems that are practical and affordable. A 5,000 psi tank, which until recently was considered at the upper limits for storage of gaseous fuel, could still take up more than ten times the volume of a gasoline tank with the same energy content. While this is acceptable for demonstration vehicles, it would not be practical for commercial vehicles. Even a 10,000 psi tank would take up seven to eight times the volume of an equivalent-energy gasoline tank or perhaps four times the volume for a tank with a comparable driving range (since the fuel cell vehicle may be twice as fuel efficient).[14] Such tanks are being demonstrated, though operation at such high pressures adds overall system complexity and requires more sophisticated materials and components, including seals and valves that are currently very expensive.

Another drawback of compressed hydrogen tanks is that they are usually cylindrical in order to ensure integrity under pressure. This undermines flexibility in vehicle design. By contrast, liquid fuel tanks can be shaped according to the needs of the vehicle. When a large cylindrical tank is used, the vehicle must be designed around the fuel tank. Also, the use of very high-pressure onboard storage would also limit refueling options to those places that invested in an expensive multi-stage compression and fueling system.

The cost of storage increases with the pressure attained. Some estimates put the cost of an 8,000 psi storage vessel at $2,100 per kilogram of capacity—more than 100 times the cost of a gasoline tank.[15] As with many storage technologies, advances in material science and economies of scale from manufacturing will reduce the cost, perhaps dramatically, while providing more flexibility in the tank design.

From a global warming perspective, compressing 1 kg of hydrogen into 10,000 psi onboard tanks could take 5 kWh or more of energy.[16] Again, since that electricity would come from the U.S. electric grid, hydrogen compression to 10,000 psi alone would for the foreseeable future release some 7 pounds of CO_2 into the atmosphere. This is one-third of the emissions resulting from burning a gallon of gasoline.

Metal Hydrides

Metal hydrides include several classes of hydrogen-containing compounds that could offer a promising means of hydrogen storage. Hydrogen is chemically bonded to one or more metals and is released through a catalyzed reaction or through heating. Hydrides can be stored in solid form or in a water-based solution. After a hydride has released its hydrogen, a by-product remains in the fuel tank to be either replenished or disposed of.

Hydrides are classified as either reversible or irreversible. Reversible hydrides act similarly to sponges, soaking up hydrogen compactly. They are generally solids—alloys or intermetallic compounds—that release hydrogen under specific pressures and temperatures. These can be replenished by adding pure hydrogen at a filling station. Irreversible hydrides are compounds that undergo reactions with other reagents, such as water, producing a by-product. You cannot simply reverse the process and restore them by supplying hydrogen; the by-product may well have to be shipped to a chemical-processing plant.

Researchers continue to struggle with suboptimal hydrogen release (release of only a portion of stored hydrogen) and refueling issues. Reversible hydrides are refueled through supply of pure hydrogen to the original alloy, and this may take much more than five minutes. Just as some hydrides are slow to absorb hydrogen in fueling, others are slow to release it during use. The chemical processes underlying irreversible hydrides can be quite energy-intensive.[17]

The major advantage of metal hydrides is their ability to contain a large amount of hydrogen in a small volume. A metal hydride tank might carry 5 kg of hydrogen in one-third the volume of a 5,000 psi tank. Moreover, hydride tanks can be designed in a variety of shapes, simplifying overall vehicle design.

Hydrides are, on the other hand, heavy. Given the actual amount of hydrogen released, the weight of the container, and the weight of other equipment, the storage capacity may be less than 2 percent by weight—each 1 kg carried may require 50 kg or more of tank, so a tank carrying 6 kg of hydrogen might weigh more than 300 kg.[18] This weight reduces fuel efficiency, which is one of the main points of having a hydrogen vehicle in the first place. Many hydrides have a theoretical capacity to store a higher percentage of hydrogen by weight; so they remain a major focus of research.[19]

Novel Storage

Novel means of storing hydrogen are the subject of intensive research because of the problems with the methods described earlier. In particular, there has been considerable experimentation with hydrogen storage in fullerenes ("buckyballs" or "carbon nanotubes"), which are microscopic structures fashioned out of carbon. Recent research indicates that there is potential for buckyball storage through a combination of chemical and physical containment but at very high temperature and pressure.[20] These novel strategies will require considerable research and development (R&D) before they can be the basis of a commercial product.

It is not clear that one storage technology will dominate the market. No storage tank technology at present has all the attributes required for a successful commercial product: compact, lightweight, safe, inexpensive, and easily filled. The 2001 study by the California Fuel Cell Partnership (CaFCP) on commercializing fuel cell vehicles concluded, "The compressed gas option is already well developed for general use despite its inherent drawbacks, LH$_2$ [liquid hydrogen] is usable but not widely considered practical, and hydrides are longer-term prospects."[21] A 2003 report by the National Research Council noted that compressed hydrogen storage at 5,000 to 10,000 psi "is costly not only for the storage canisters onboard vehicles but also for the compressors and electricity required for compression at refueling stations."[22]

It could well be that various technologies will be used, according to the application. Cars, sport utility vehicles, vans, buses, and heavy trucks all have different needs, and these needs also vary by application (such as city-owned fleet trucks versus long-haul trucks). Some vehicle owners will place a premium on volume, others on weight, and others on cost or some other factor.

Finally, while compressed gas appears to be winning the battle for near-term demonstration vehicles, the path to commercialization of any major new technology is a long one. In May 2003, Toyota recalled all six of its hydrogen-powered fuel cell vehicles when a leak was discovered in the fuel tank of one of the cars leased to Japan's Ministry of the Environment.[23] The leak was found when "a driver at the environment ministry heard a strange noise in the car when he was filling up the hydrogen tank."[24] The problem was quickly identified and was fixed a few weeks later.

Nationwide Infrastructure

If hydrogen is produced at local fueling stations from natural gas (or electricity), then no significant new nationwide infrastructure for hydrogen delivery will be required. A pure hydrogen economy requires that hydrogen be generated from CO_2-free sources, which, as discussed in the previous chapter, would almost certainly require centralized hydrogen production closer to giant wind farms or at coal or biomass gasification power plants where CO_2 is extracted for ultimate sequestration. That will require some way of delivering hydrogen to thousands, and ultimately tens of thousands, of local fueling stations. The options most widely discussed are tankers delivering liquid hydrogen, hydrogen gas pipelines, and trailer trucks delivering high-pressure canisters.

Tanker trucks carrying liquefied hydrogen are commonly used to deliver hydrogen today. This is, as mentioned, the method NASA uses, and it is an increasingly popular delivery strategy in Europe and North America to "more effectively supply many distributed consumers with moderate hydrogen demands," as one analysis put it, because it is "currently cheaper than small on-site hydrogen generation systems, and assures higher purity and availability."[25] Most current users require these moderate amounts of hydrogen for high-value-added industrial and manufacturing processes, so their top priority is the purest, least expensive, and most flexible means of ensured delivery. How much energy is consumed and how much pollution is released in making the hydrogen generally are not top-priority issues for these users. But for hydrogen used as a transportation fuel, energy and pollution are of the highest concern.

While some analysis has suggested that liquid tanker trucks could be the least expensive delivery option in the near term—and carry ten times the amount of hydrogen transported by trucks carrying compressed hydrogen canisters—this approach is undesirable for large-scale use. Liquefaction has a sky-high energy cost, some 40 percent of the usable energy in hydrogen. Moreover, few automakers are pursuing onboard storage with liquid hydrogen. So, after delivery, the fueling station would still have to use an energy-intensive pressurization system, which would consume another 10 to 15 percent of the usable energy in the hydrogen. This might mean that storage and transport alone might require as much as 50 percent of the energy in the hydrogen delivered, most likely in the form of electricity. Barring a dramatic effort to reduce U.S. greenhouse gas emissions in the electric utility sector, this approach by itself might result in as much CO_2 emissions as would be avoided by replacing gasoline with hydrogen produced from zero-carbon sources. For liquefaction to be a viable option, a significantly less energy-intensive process needs to be developed.

Pipelines are also used for delivering hydrogen. Today, several thousand miles of hydrogen pipelines are in use around the world, of which several hundred miles are in the United States.[26] They are short and located in industrial areas for large users. The longest is about 250 miles long, running from Antwerp to Normandy. It operates at 100 atmospheres of pressure (roughly 1,500 psi).

Pipelines might well be the least expensive option for delivery of large quantities of hydrogen. Pipelines are the main option for transporting refined petroleum products across the country "because they are at least an order of magnitude cheaper than rail, barge, or road alternatives."[27] The United States has some 200,000 miles of interstate pipeline carrying petroleum in various forms.[28] We also have more than 200,000 miles of interstate natural gas pipelines.[29]

Hydrogen pipelines are very expensive, in part because they are carrying a fuel that is very diffuse and prone to leaks. Hydrogen is also highly reactive. It can cause many metals, including steel, to become brittle over time. Large-scale interstate hydrogen pipelines (nine to fourteen inches in diameter) have a capital cost of $1 million per mile or more, whereas smaller pipelines for local distribution might cost half of that.[30]

Clearly, hydrogen infrastructure designers would be very strategic about where pipelines are placed to minimize total construction costs. The problem in the near term (the first two decades of this century) is that it is far from clear where hydrogen will be produced and far from certain that hydrogen will succeed in the marketplace. Siting major new oil and gas pipelines is often politically and environmentally controversial, but in both cases we know where the source of the oil and gas is, and typically we know with some confidence an end point with high demand for oil and gas, even if the path the pipeline might take between them is less certain. We have very little idea today which hydrogen generation processes will win in the marketplace over the next few decades. Nor can we know what political pressures will be brought to bear to favor one generation technology over another, or whether global warming concerns will be so great as to override all other considerations. Nor is it clear that pure hydrogen fuel cell vehicles will be economically competitive with future high-efficiency vehicles, perhaps running on other low-carbon fuels.

All this uncertainty makes it unlikely that anyone would commit to spending billions of dollars on hydrogen pipelines before there are very high hydrogen flow rates with other means of transport, and before the winners and losers have been determined at both the production end and the vehicle end of the marketplace. In short, for all their virtues, pipelines are not likely to be the main hydrogen transport means over at least the next two decades.

Trailers carrying compressed hydrogen canisters are a flexible means of delivery probably best suited for early market introduction of hydrogen. This is, however, a relatively expensive delivery method because hydrogen has such a low energy density, so even with relatively high-pressure storage, very little hydrogen is actually being delivered. Current tube or canister trailers deliver less than 300 kg of hydrogen to a customer, perhaps enough to fully fill sixty fuel cell cars.[31] An analysis in 2003 by Ulf Bossel and Baldur Eliasson of hydrogen delivery in improved high-pressure canisters estimated that a 40 metric ton truck would deliver only about 400 kg of hydrogen into on-site high-pressure storage.[32]

Perhaps the simplest option would be to leave the canisters or the tube trailers at the point of delivery. This strategy avoids the need for building costly on-site high-pressure storage, but it may not decrease overall costs because the high-pressure canisters are likely to be expensive and, in this option, are not very efficiently used: The delivery fleet needs far more tube trailers and canisters because they're all doing double duty for both distribution and storage. When they are empty, they are picked up to be refilled.[33]

In any case, from an energy perspective, this delivery approach is exceedingly wasteful. A 40,000 kg truck is being used to deliver a mere 400 kg of hydrogen. More than 39 unused metric tons, a tremendous amount of dead weight, would be hauled around our highways, burning up diesel fuel. Compare that with the fact that the same size of truck carrying gasoline delivers some 26 metric tons of fuel (10,000 gallons), enough to fill perhaps 800 cars.

Bossel and Eliasson draw two striking conclusions from their analysis. First, the energy consumed by tanker truck delivery itself can be a very high fraction of the total hydrogen energy delivered. For a delivery distance of 150 miles, the delivery energy equals nearly 20 percent of the usable energy in the hydrogen delivered. For 300 miles, the energy ratio approaches 40 percent.

Second, a huge number of trucks would be needed if they were the primary delivery means. They conclude, "It would take 15 tube-trailer hydrogen trucks to serve the same number of vehicles that are nowadays energized by a single 26-ton gasoline truck." They go further and note that nowadays "about one in 100 trucks is a gasoline or diesel tanker," or 1 percent, so replacing liquid fuels with hydrogen transported by tube truck could mean that more than 10 percent of the trucks on the road would be transporting hydrogen. This would raise serious logistical and safety concerns.

Technology may well present superior options in the future—and there is considerable R&D into each and every one of these storage and transportation technologies—but these first-approximation calculations must be thoroughly understood to see how far we are from a practical transition to a hydrogen economy. There do not seem to be any particularly attractive near-term options for delivering significant quantities of hydrogen to fueling stations, which would seem to support the view that we should be looking at generating hydrogen at those fueling stations. Before addressing that issue, let's look at the issue of hydrogen safety.

Hydrogen Safety

As noted in the previous chapter, hydrogen is widely used in industry today, and we have decades of experience in generating, storing, and transporting it. It has a good safety record in industrial settings.

Hydrogen's bad reputation as a transportation fuel stems in part from the *Hindenburg* disaster.[34] The largest zeppelin ever built, with a maximum speed of more than eighty miles per hour, the *Hindenburg* was an 800 foot airship kept aloft by large bags filled with hydrogen. Passengers rode in a luxurious gondola underneath. At the time of its fateful voyage in May 1937, it had made numerous uneventful trips to Brazil and North America.

Traveling from Frankfurt, Germany, the *Hindenburg* arrived in Lakehurst, New Jersey, on May 6, 1937, during an electrical storm. Suddenly, the zeppelin burst into brilliant red and orange flames, and about thirty seconds later it crashed into the ground. It was one of the most spectacular disasters of the twentieth century. Of the ninety-seven people on board, thirty-five died. Investigations by the American and German governments concluded that the cause was a flammable mixture of hydrogen and air that supposedly formed under the dirigible's cover. Low-pressure hydrogen flames are, however, virtually invisible.

Former NASA scientist Addison Bain studied the tragedy for nine years, reviewing original documents, talking to eyewitnesses, and performing laboratory experiments. He and German scientist Ulrich Schmidtchen have made a strong case that the cause of the accident was the paint on the *Hindenburg*'s outer cover, which was a highly flammable mixture similar in composition to rocket fuel, igniting during the electrical storm. Indeed, the German archives reveal that an electrical engineer named Otto Beyersdorff involved in the original investigation had concluded, "The actual cause of the fire was the extremely easy flammability of the covering material brought about by the discharges of an electrostatic nature." Nonetheless, hydrogen no doubt fed the *Hindenburg*'s fire and contributed to the speed with which the giant airship burned.[35]

The fact that hydrogen probably was not the source of the *Hin-*

denburg accident does not, of course, prove that hydrogen is a safe fuel. Hydrogen does have some advantages over liquid fuels such as gasoline. When a gasoline tank leaks or bursts, the gasoline can pool, creating a risk that any spark might start a fire, or it can splatter, posing a great risk of spreading an existing fire. Hydrogen, however, will escape quickly into the atmosphere as a very diffuse gas. Also, hydrogen gas is non-toxic.

On the other hand, hydrogen has its own major safety issues. It is flammable over a wide range of concentrations and has a minimum ignition energy twenty times smaller than that of natural gas or gasoline: "Operation of electronic devices (cell phones) can cause ignition" and "common static (sliding over a car seat) is about ten times what is needed to ignite hydrogen," as explained in a 2002 report by Arthur D. Little Inc.[36] Even "electrical storms several miles away" can generate sufficient static electricity to ignite hydrogen, noted James Hansel, a senior safety engineer for Air Products and Chemicals Inc., a major global supplier of hydrogen, in a 1998 presentation to Ford Motor Company.[37]

Hence, leaks pose a significant fire hazard. At the same time, with its tiny molecule, hydrogen is one of the most leak-prone of gases, and, in the case of liquid hydrogen, a small fraction will boil off every day. Hydrogen is subject to strict and cumbersome codes and standards, especially where it is used in an enclosed space and a leak might create a growing bubble of hydrogen gas. Hydrogen expert Dale Simbeck noted: "Current codes for H_2 storage and utilization are onerous. For example, in the United States, the current National Fire Protection Association (NFPA) code 50 A&B for H_2 does not allow underground storage, requires large clearances from all combustible materials, and requires massive ventilation to avoid very expensive explosion-proof equipment."[38]

Work is ongoing to develop less onerous codes and standards, but for now they greatly complicate hydrogen infrastructure efforts. "Fuel-cell vehicles will require modifications to garages, maintenance facilities, and on-road infrastructure [such as tunnels] that could be costly and difficult to implement," as the A. D. Little report noted. "Implementation of critical safety measures for closed public structures may pose a serious hurdle to widespread use of compressed hydrogen."[39]

To make sure we can easily detect leaks with our noses, odorants are added to flammable, odorless gases, as in the case of natural gas. But that is probably impractical with hydrogen. As Jim Campbell of Air Liquide explains, adding odorants to hydrogen "adds a contaminant that is poisonous to many fuel cell technologies," which "will add cost to the hydrogen energy picture, either in pretreatment to remove the contaminants, or in reduced service life of the affected systems."[40] Moreover, even if practical odorants could be found, they would be highly unlikely to leak and diffuse in tandem with hydrogen, the tiniest and lightest of all molecules.

The real danger exists that people might step into the nearly invisible hydrogen flame and burn themselves. According to James Hansel, "people have unknowingly walked into hydrogen flames."[41]

How can invisible leaks and flames be detected? A variety of optical, ultraviolet, and infrared sensors are available, but, Hansel notes, "fixed detectors indoors (e.g., inside some volume, compartment) or outdoors (e.g., no volume constraint), unless many are utilized, or the volume is small, *are not highly effective.*"[42] In fact, an Air Products safety document recommends, "if a leak is suspected in any part of a hydrogen system, it is good protection to hold a large square of paper before you and approach the suspected leak so that any invisible flame will strike the paper before striking you." NASA, which has decades of experience in dealing with hydrogen and access to the most sophisticated technology, also sees the limitation of sensors. In its safety guidelines for hydrogen systems, it lists four detection technologies. Three involve advanced sensors, but the fourth is decidedly low-tech: "A broom has been used for locating small hydrogen fires," since "a dry corn straw or sage grass broom easily ignites as it passes through a flame."[43]

Industrial codes and standards for hydrogen are thus strict for a good reason. Some 22 percent or more of hydrogen accidents are caused by undetected hydrogen leaks.[44] And this is "despite the standard operating procedures, protective clothing, and electronic flame gas detectors provided to the limited number of hydrogen workers," as Russell Moy, former group leader for energy storage programs at Ford Motor Company, wrote in the November 2003 issue of the *Energy Law Journal*. Moy concludes, "With this track record, it is difficult to imagine how hydrogen risks can be managed acceptably by the general public when wide-scale deployment of the safety precautions would be costly and public compliance impossible to ensure."[45]

In addition, any high-pressure storage tank presents a risk of rupture, and, as we have seen, the current trend in hydrogen storage is to use very high pressures. Hydrogen pipelines are also likely to be under very high pressure to keep the gas flowing. In the event a tank is punctured, it is critically important that the tank does not shatter and the fuel does not ignite. Storage tanks are routinely tested under a range of harsh conditions (including puncture by armor-piercing ammunition) to ensure safety. Still, gaseous hydrogen will leak through most materials over time. Therefore, storage tanks and pipelines for hydrogen must contain liners of a material (such as a polymer compound) that is impermeable to hydrogen. Metal hydride systems may have a safety advantage because they operate at relatively low pressures, typically less than 200 psi. In a crash, these systems wouldn't tend to release hydrogen, since it is chemically bonded to the hydrides in a solid state.[46]

Hydrogen is also highly reactive. It can cause many metals, including steel, to become brittle. This raises the risk of cracking or fissuring, which can result in system failure ("potentially catastrophic failure of pipelines"). Even low-permeability liners will allow some hydrogen through to the tank material; therefore, storage tanks must be resistant to hydrogen embrittlement. Air Liquide's Jim Campbell notes that "higher strength materials are more susceptible to hydrogen embrittlement."[47]

If a large hydrogen infrastructure is built without major new safeguards, we can expect to have more and more hydrogen accidents. In May 2003, for instance, a tanker truck carrying compressed hydrogen caught fire in California, "sending a hissing flame as high as 60 feet into the air," according to news reports.[48] Firefighters were able to put out the blaze within a few hours and avert a potentially dangerous explosion. Damage was minimal, beyond a loss of about one-third of the hydrogen carried by the truck. Such accidents are routine with existing energy fuels, such as gasoline and natural gas, and as such they do not represent fatal flaws in a hydrogen economy. On the other hand, for a new fuel trying to break into the marketplace, especially one linked to a famous fiery disaster, any accidents will slow efforts to change safety perceptions and, more important, to change onerous safety regulations. Thus, the critical path for a hydrogen economy must include major innovations in safety.

Tough Infrastructure Choices

Having reviewed some of the key technical issues, let's take another, more strategic look at the main infrastructure choices that would support a major shift to fuel cell vehicles. The nature of the hydrogen infrastructure, as noted earlier, will be driven by where the hydrogen is produced and in what form it is stored on board the hydrogen vehicle. Each of the major choices—onboard hydrogen production, centralized production, and production at fueling stations—faces serious technological and economic barriers.

Onboard hydrogen production will require significant R&D to solve the cost and performance issues. Reforming either methanol or gasoline into hydrogen on board a car is likely to be less efficient than stationary reforming. Onboard reformers produce less pure hydrogen streams, which reduces the fuel cell's efficiency. Overall efficiency for gasoline and methanol fuel cell vehicles may therefore be much lower than for hydrogen fuel cell vehicles.

In the case of onboard reforming of gasoline to hydrogen, both the global warming and energy security benefits appear likely to be rather modest. A 2003 "Comparative Assessment of Fuel Cell Cars" by the Massachusetts Institute of Technology found that the life-cycle energy consumption and greenhouse gas emissions of a future gasoline–fuel cell vehicle would actually be *greater* than that of a future hybrid gasoline-electric vehicle.[49] This suggests that the primary potential benefits of onboard gasoline reforming are (1) to capture economies of scale by accelerating commercialization of PEM fuel cells and (2) to avoid the hydrogen infrastructure problems laid out in this chapter, thus directly solving the chicken-and-egg problem. However, given that gasoline–fuel cell vehicles will probably cost far more than hybrid gasoline-electric vehicles, this benefit may ultimately prove too insignificant for governments, businesses, and individuals to justify the investment.[50]

Onboard methanol reformers appear likely to be even less efficient than gasoline reformers. Also, for the foreseeable future, any significant increase in methanol production will most likely come from overseas natural gas, in a process that itself is comparable in efficiency to producing hydrogen from natural gas. So the energy and greenhouse gas benefits also appear likely to be rather modest.[51] The existing methanol infrastructure could handle a not-too-ambitious introduction of methanol fuel cell vehicles at a very low cost, perhaps less than $50 per car.[52] But any significant use of methanol as a transportation fuel would require a major investment in fuel production and delivery, which may well be as expensive as a hydrogen infrastructure.[53]

The biggest obstacle for methanol is probably the widely held view that hydrogen is the inevitable fuel—because hydrogen can potentially be made from a variety of zero-carbon sources and because pure hydrogen fuel cell vehicles are likely to be much more efficient. This raises the specter that any significant investment in a methanol infrastructure might become "stranded," or rendered obsolete, like Betamax videocassette recorders. A 2000 analysis by Princeton University's Center for Energy and Environmental Studies concluded, "A serious disadvantage of this [methanol] scenario is that the fuel infrastructure would have to be changed twice (from gasoline to methanol, then from methanol to hydrogen)."[54]

For the fueling companies that would have to make the double investment, the choice is not an academic concern. In September 2000, the Royal Dutch/Shell Group, which has invested heavily in hydrogen, raised this very issue. Sir Mark Moody-Stuart, then chairman of Shell, said at a hydrogen conference, "Some companies have suggested the use of other transition fuels such as onboard methanol-to-hydrogen conversion. Such fuels would require significant infrastructure investments, which would be difficult to justify for what could be short-lived solutions."[55]

Given the enormous cost and logistical complexity of replacing the existing gasoline infrastructure, we must strive to ensure that any major near-term investments are compatible with long-term goals and solutions. One guiding principle of the transition to a hydrogen economy should be to *minimize the likelihood of stranded investments*. Simply put: Let's do this only once.

The ultimate goal is a pure hydrogen economy, one based on zero-CO_2 sources of hydrogen. We must therefore avoid devoting significant resources to an infrastructure that is incompatible with that goal. We should withhold judgment on methanol as the primary hydrogen carrier until we have the following:

- An affordable and practical direct methanol fuel cell.
- An affordable and practical means of generating large quantities of methanol from renewable sources.

These targets will doubtless take more than a decade of R&D.

Centralized production of hydrogen almost certainly would deliver less expensive hydrogen than production at local fueling stations, especially where large volumes are needed. It is also the approach needed in a pure hydrogen economy, one with carbon-free generation, which is our ultimate goal. On the other hand, as we have seen, this option "will be difficult to implement in the near term on a large scale because of its infrastructure challenges," as an Arthur D. Little analysis from 2000 put it.[56] So-called resource-centered hydrogen production near large energy resources (such as sources of natural gas) may be straightforward, but the relatively inexpensive means of generation—natural gas, coal gasification—are not carbon-free, and the carbon-free generation methods—such as wind power and biomass—are relatively expensive.

Moreover, hydrogen delivery is problematic with existing technology, as noted earlier. The least expensive option, tanker trucks carrying liquefied hydrogen, is extremely energy-intensive and would negate much of the benefit of a hydrogen economy. Pipelines may be the least energy-intensive option, but they will be the most expensive until there are very high rates of utilization. The high up-front capital cost means that someone must take a big gamble that the pipeline is being put in the right place and will achieve high utilization. It seems unlikely that anyone—including the federal government—would take such a gamble until hydrogen vehicles constitute a large fraction of the vehicles on the road, and so this represents a longer-term option. Trucks carrying compressed hydrogen canisters make sense for the very initial introduction of hydrogen into the marketplace, but from an energy, economic, and logistical perspective, this method is probably impractical for sustaining any significant market penetration of hydrogen as a transportation fuel.

Finally, a 2002 study by Argonne National Laboratory examining different infrastructure options calculated that infrastructure costs for fueling about 40 percent of the vehicles on the road with resource-centered hydrogen production would be $600 billion, an enormous cost. Not surprisingly, a great many analysts and advocates dismiss centralized hydrogen production as impractical for the foreseeable future.

Production of hydrogen at local fueling stations is the strategy advocated in particular by those who want to deploy hydrogen vehicles quickly, such as the Rocky Mountain Institute (RMI). In the RMI scenario, the hydrogen would most likely be generated from small steam methane reformers (SMRs); electrolysis would most likely be more expensive and far more polluting (see Chapter 4). This seems viable for limited demonstrations and pilot projects, but is both impractical and unwise for large-scale application for several reasons.

The first reason is that the up-front cost is higher with technology that is commercial today. In this scenario, a fueling station would need an SMR system, a hydrogen purification unit, a multistage hydrogen compressor (since a high-pressure tank is the only viable near-term system for storage on board a fuel cell vehicle), a system to fuel cars, a significant amount of on-site high-pressure storage, and various types of control equipment. These components, such as SMRs, do exist, though some are only in early stages. SMRs typically have economies of scale—large plants are relatively cheaper to build than small ones. The Argonne study assumed 50 percent cost reduction in compressors over current technology and still found that infrastructure costs for fueling 40 percent of the vehicles on the road would exceed $600 billion. There are studies that find dramatically lower costs, but they all rely on technology not commercial as of 2003 and cost projections that are quite optimistic.

Second, once demand becomes significant, the cost of the delivered hydrogen itself in this option is also higher than for centralized production. Not only are the small SMRs and compressors typically more expensive and less efficient than larger units, but they also will very likely involve a much higher price for the electricity and gas to run them. Centralized units can much more easily time their major electricity consumption for compression to coincide with off-peak rates than fueling stations can (as noted in the previous chapter). Dale Simbeck and Elaine Chang, in their scoping analysis of different hydrogen supply options for the National Renewable Energy Laboratory, found that forecourt hydrogen production at fueling stations by electrolysis from grid power was by far the most expensive option, with a cost of $12/kg, and forecourt natural gas production was $4.40/kg.

Again, those studies that find dramatically lower costs rely on technology that is not commercial as of 2003. A 2001 study for the California Fuel Cell Partnership concluded that "at least one cost-effective on-site fuel processing and vending technology should be developed and introduced before [fuel cell vehicle] market introduction begins."[57] Whoever succeeds in building an inexpensive small SMR will make a great deal of money selling that system to the growing number of industrial hydrogen users.

Third, "the risk of stranded investment is significant, since much of an initial compressed hydrogen station infrastructure could not be converted later if either a non-compression hydrogen storage method or a liquid fuel such as a gasoline-ethanol combination proved superior for FCVs [fuel cell vehicles]," as noted by the study, which examines forecourt production in great detail.[58] Most of the investment would also very likely be stranded once the ultimate transition to a pure hydrogen economy was made, since that would almost certainly rely on centralized production and would not make use of small SMRs. Moreover, if the solution to the chicken-and-egg problem is to build the infrastructure first with a massive government subsidy, then it is possible the entire investment would be stranded in the scenario in which FCVs simply never achieve the combination of popularity, cost, and performance needed to triumph in the marketplace.

The study notes, "A substantial risk premium may thus be applied by potential hydrogen infrastructure investors." In this analysis, it takes ten years for an investment in infrastructure to achieve a positive cash flow, far too long for the vast majority of investors, and, to achieve this result, "significant technological advances will be required in reformers and electrolyzers, compressors, and overall systems integration as well as mass production methods for that equipment." Also, even a small excise tax on hydrogen (to make up for the revenue lost from gasoline taxes) appears to delay positive cash flow indefinitely.[59]

Fourth, *natural gas is the wrong fuel on which to base a hydrogen-based transportation system.* I will address this point in more detail in the next two chapters, but the key points are as follows:

- A large fraction of new U.S. natural gas consumption will probably be supplied from overseas. While those sources are no doubt more secure than some of our sources for oil, replacing one import with another does not move us in the direction of energy independence.
- Natural gas can be used far more efficiently to generate electricity or to cogenerate electricity and steam than it can be to generate hydrogen for use in cars. Using natural gas to generate significant quantities of hydrogen for transportation would, for the foreseeable future, undermine efforts to combat global warming (see Chapter 8).

These conclusions do not mean that we cannot start generating hydrogen from natural gas—most U.S. hydrogen currently comes from natural gas. They do show that it would be unwise to build hundreds, let alone thousands, of local fueling stations based on SMR (or based on any technology not easily adaptable to delivery of zero-carbon hydrogen).

CHAPTER 6

The Long Road to Commercialization
of Fuel Cell Vehicles

H ydrogen fuel cell cars are unlikely to achieve significant market penetration in this country by 2030. As discussed in the previous chapter, the infrastructure costs alone are likely to run into the hundreds of billions of dollars. Is there a way to jump-start the transition? Could fuel cell cars really do double duty as mobile power plants, providing electricity to our homes and offices, as energy experts such as Amory Lovins have suggested? These are the questions I examine in this chapter.

Right now, we still have a lot of research, development, and analysis to do in all aspects of the hydrogen economy—production, storage, transport, fueling, safety, and, most of all, use. Consider the vehicles. In December 2002, Yozo Kami, Honda's engineer in charge of fuel cells, said it would take at least ten years to bring the sticker price of a fuel cell vehicle down to $100,000—and as of 2003 Honda had one of the least expensive prototype fuel cell vehicles (FCVs).[1] A major analysis for the U.S. Department of Energy (DOE) concluded in 2002 that even using relatively optimistic assumptions about technological improvement, "factory costs of future FCVs would likely be 40–60% higher than [for] conventional vehicles."[2]

On the fueling side, we also have a long way to go. Don Huberts, former chief executive officer of Shell Hydrogen and former chair of the California Fuel Cell Partnership (caFCP), bluntly told the *Fuel Cell Industry Report* in January 2003, "At the end of the day, hydrogen and other alternative fuels will be three to four times as expensive as oil-based products, and if no one wants to pay for that, we can't make those fuels."[3] This is especially true for hydrogen from zero-carbon sources, such as renewable energy.[4]

Let's explore some options that have been proposed to accelerate the commercialization of hydrogen vehicles.

Hydrogen-Powered Internal Combustion Engines

Introducing hydrogen-powered internal combustion engine (ICE) vehicles is one strategy advocated by experts such as S. David Freeman, chair of the California Consumer Power and Conservation Financing Authority. Freeman said in June 2003, "I am troubled by the propaganda and hype of fuel cells. . . . We don't have to wait for the fuel cell. We can't afford to wait for the fuel cell. Indeed, if we start to make hydrogen-burning cars immediately, we will bring on the infrastructure that will make it possible for fuel cell cars to come in sooner."[5]

An internal combustion engine can run on hydrogen with relatively inexpensive modifications. A number of automakers, including the Ford Motor Company and BMW, are planning to introduce hydrogen ICE cars. They have the advantages over gasoline ICEs of very low emissions of urban air pollutants and the possibility of running on fuels that are not imported. Introduction of hydrogen ICEs avoids the time and cost associated with proton exchange membrane (PEM) fuel cells, thus potentially speeding up the transition to a hydrogen economy.

This strategy, for all its compelling logic, does not solve a number of the fundamental issues that are delaying the transition. It does not address the problem of the high cost of hydrogen or the life-cycle cost of the vehicle. Hydrogen ICEs are likely to be far less efficient than fuel cell vehicles and perhaps only 25 percent more efficient than gasoline ICEs. Therefore, they are likely to have a reduced driving range because of the difficulty of storing large volumes of hydrogen on board.[6] Vehicle owners would directly experience the high price of hydrogen. As a result, annual vehicle ownership costs for mid-sized hydrogen ICE vehicles would be 30 percent higher than for current gasoline ICE vehicles (and only slightly lower than for fuel cell vehicles), according to a 2002 analysis by Arthur D. Little Inc.[7]

Moreover, because of the inevitable losses in generating hydrogen (from natural gas or electricity, for instance) and because of the energy consumed in compressing hydrogen for storage, the "well-to-wheel" energy use of a hydrogen ICE vehicle may actually be higher than that of a gasoline ICE vehicle. A 2002 analysis of ten different alternative fuel vehicles (AFVs) found that ICEs running on hydrogen from natural gas had the lowest overall efficiency on a life-cycle (well-to-wheel) basis.[8] Running an ICE car on hydrogen from natural gas would probably not save any greenhouse gas emissions compared with running a gasoline internal combustion engine car and would *increase* emissions compared to a hybrid gasoline-electric car. If mitigating global warming is your goal, hydrogen internal combustion engine cars are not a viable strategy for the foreseeable future.

Hydrogen ICEs probably cannot accelerate the transition to a hydrogen economy. They do not solve the chicken-and-egg problem. It seems unlikely that the public would be willing to pay the extra cost of owning and operating such vehicles, given the relatively modest benefits they provide. It seems unlikely that governments or companies would be willing to build an expensive infrastructure for such vehicles before seeing whether the public would embrace them.

Fleets

Introducing AFVs through fleets is another widely discussed strategy for accelerating commercialization. As discussed in the previous chapter, it was the primary strategy that the DOE adopted in the 1990s to meet the AFV goals of the Energy Policy Act of 1992. Vehicle fleets are an attractive prospect for AFV commercialization because they are typically driven twice as far in a given year as household vehicles. Therefore, fleet owners, in comparison to general consumers, could achieve higher energy and emissions benefits by using AFVs.[9]

Fleet vehicles represent about one-quarter of all U.S. light-duty vehicle sales, and "relatively few decisionmakers control a disproportionately large number of vehicles," according to a 1998 study by the University of California, Davis. Finally, "Many fleet vehicles have fixed daily routes and are regularly fueled at one location," which is perhaps the primary reason why they are of interest, since far less infrastructure might be needed to support fleet-based fuel cell vehicles.[10]

The DOE's lack of success in promoting AFVs to fleet owners, however, suggests that the picture is more complicated. The 1998 study, "Myths Regarding Alternative Fuel Vehicle Demand by Light-Duty Vehicle Fleets," based on interviews and focus groups with dozens of fleet operators and managers as well as a two-part survey of more than 2,700 California fleets, helps reveal why AFVs have had so much difficulty in penetrating fleets. The study does not paint a wholly negative picture, but it notes two reasons why central fueling may actually deter a fleet owner from purchasing certain AFVs. First, "light-duty fleets that centrally refuel on-site typically do so because they have high travel demands and, therefore, can significantly reduce fuel costs by purchasing petroleum in bulk." Yet hydrogen is currently three to four times more expensive than gasoline on an equivalent energy basis, and fuel cell vehicles in particular are unlikely to have lower fueling or life-cycle costs for many decades. Second, "high travel demands preclude the use of most AFVs, which have shorter ranges and limited refueling opportunities." That is, current gasoline-powered vehicles have a long driving range and can easily get to their large central fueling stations for less cost (or, if necessary, can partially refuel at a commercial gas station). But AFVs, with shorter driving ranges and fewer fueling stations, cannot.[11]

Also, while 78 percent of public fleets surveyed used central refueling, only 34 percent of business fleets did, and few of those did so exclusively: "Most fleets that centrally refuel rely on outside sources for at least 15% of their refueling needs."[12] Ironically, the very fact that

AFV fleet mandates typically apply only to fleets that are centrally refueled appears to be accelerating a trend away from central refueling—as fleet owners seek to avoid being forced to purchase AFVs. Also, the U.S. Environmental Protection Agency's regulations and liability issues concerning fuel leakage led to a significant decrease in the number of underground fuel storage tanks between 1989 and 1999. Thus, the business trend is away from central refueling.

Governments can more directly mandate purchase of AFVs with public fleets than they can with private fleets. On the other hand, my colleague Brian Castelli and I have years of personal experience at the DOE in trying to jawbone government agencies into spending their scarce dollars on AFVs that were required by the Energy Policy Act and by President Bill Clinton's executive orders. The AFVs were typically far more expensive than the cars and trucks the agencies had been purchasing, which forced them into tough choices, since their budgets for vehicle purchase were fixed by Congress.

In the current era of high federal budget deficits and tight state budgets, it will be interesting to see whether taxpayers and legislators are willing to spend scarce dollars to subsidize hydrogen vehicles and hydrogen fuel. In 2002 and 2003, we have already seen a number of states dip into funds that were dedicated to promote clean energy, using them for other purposes and creating a very bad precedent for the future. There is little evidence that targeting fleets can avoid the chicken-and-egg problem. A 2000 report on AFVs by the U.S. General Accounting Office (GAO) concluded, "Officials from federal agencies and state governments who administer vehicle fleets cited the lack of a refueling infrastructure more than any other impediment to using alternative fuels."[13]

Ultimately, the question is whether fleets represent a way to jump-start the consumer market for hydrogen fuel cell vehicles. According to the GAO report, "several fleet managers and representatives of the automobile industry acknowledge it is unlikely that usage of alternative fuel vehicles by these fleets will convince the general public to buy them."[14] Thus, while fleets remain an important entry market for fuel cell vehicles, we will probably need a different strategy to achieve broad commercialization.

Fuel Cell Cars as Mobile Utilities

Can we address the high cost of fuel cell vehicles by somehow finding extra economic value in those vehicles when they are not being
used, thereby reducing their initial cost and accelerating market
introduction? Most of us use our cars for less than 10 percent of the
twenty-four hours in the day, and when we are not using our cars
they are typically parked near places that consume a great deal of
electricity—our homes, offices, and factories. If every car on the
road were a fuel cell vehicle, the combined electricity generation
capacity would be several times greater than current U.S. electricity
generation.

Intrigued, many analysts have proposed plugging parked fuel
cell cars into the grid to generate power so that the transportation
fuel cell would do double duty as a stationary fuel cell. In his 2002
book *The Hydrogen Economy*, Jeremy Rifkin comments, "If just a
small percentage of drivers used their vehicles as power plants to
sell energy back to the grid, most of the power plants in the country would be eliminated altogether."[15] Yet *the possibility is remote that
any significant fraction of U.S. electricity will be generated this way
within the next three decades. We can blame our usual suspects—
technology, cost, and infrastructure.*

Consider the issue of electricity price and greenhouse gas emissions. For the next two decades at least, the dominant source of
hydrogen will most likely be reformed natural gas. Yet generating
electricity with a car's low-temperature fuel cell using that reformed
gas would cost $0.19 per kilowatt-hour (two and a half times the
national average price for electricity), according to a 2001 study
conducted for the state of California and city of Los Angeles.[16] The
process also produces 50 percent more carbon dioxide (CO_2) emissions than the best natural gas conversion technology in use today,
combined cycle natural gas turbines. And it would produce twice
the CO_2 emissions of high-temperature fuel cells used with gas turbines, which might well be producing significant electricity by 2020.

As discussed previously, low-temperature fuel cells running on
natural gas are not very efficient at generating electricity. A station-

ary fuel cell system achieves both high efficiency and greenhouse gas reductions only by co-generating heat, but cogeneration is probably not very practical for your car. Hooking up a vehicle to the electric grid will be complicated enough and will most likely require expensive electronic hardware. Extracting useful exhaust heat from the fuel cell and temporarily connecting it to either your heating system or your hot-water system is logistically even more complicated, involving new ductwork and possibly heat exchangers. This is particularly true for existing buildings, in which parking is often not adjacent to heating units. And while most homes could probably find productive use for the heat from a 1 kilowatt (kW) fuel cell, a car will probably have a 60–80 kW fuel cell.[17]

In general, home electricity generation with either a stationary or a mobile fuel cell seems very unlikely to provide synergistic cost savings that can jump-start commercialization.[18] Another key question is, how will hydrogen get to your home or office to power the fuel cell after your car's onboard hydrogen is consumed? For the relatively small amounts of hydrogen that a fuel cell car would need, delivering hydrogen is likely to be prohibitively expensive per kilogram. Yet it would probably be even more expensive to generate hydrogen on-site. As we have seen, hydrogen generation and purification units are expensive, local electricity and natural gas prices are much higher than for larger users, and on-site storage raises serious safety issues for an odorless, leak-prone, and highly ignitable gas that burns invisibly.

In the longer term, the goal is to produce zero-carbon hydrogen from renewable sources, most likely through electrolysis of water. Yet it would hardly make much sense to generate electricity from renewable sources, then generate hydrogen from that electricity using an expensive and energy-intensive electrolyzer, ship the hydrogen over long distances (consuming more energy), and then use that hydrogen to generate electricity again with low-temperature fuel cells in cars. More than half of the original energy would be lost in this complicated process (see Chapter 8).

Finally, transportation fuel cells are typically being designed for about 4,000 hours of use, which might give a car a ten-year lifetime or longer because they're used only a small percentage of the time. But 4,000 hours represents less than half a year of continuous use for generating electricity, which is why fuel cells for power plants typically are designed for 40,000 hours or more. To meet the stringent cost goals required for proton exchange membrane (PEM) fuel cells for transportation—well under \$100/kW—the membranes are being made very thin and less durable than for stationary applications. As of 2003, achieving a lifetime of even 1,000 hours has proven difficult. Meeting the cost and performance targets for PEM cars will probably prove so difficult that we can't bank on the same hardware being able to meet the performance characteristics of a power plant.

So it seems unlikely that transportation fuel cells will be generating much of the nation's electricity on a continuous basis in the coming decades. It is, however, possible to extract value from intermittent electricity generation. For instance, during the hottest days of summer, when air-conditioning electricity demand is at its highest, electricity generation costs can be quite high. This peak power generation could represent a source of revenue, although only in places where peak prices are exceedingly high. The California study suggested that the largest potential revenue stream that a vehicle owner might be able to extract is for so-called spinning reserves, which, as one analysis explains, "are contracts for generating capacity that is up and running, and is synchronized with the power line." When called upon, a spinning reserve "must ramp up to its full output within 10 minutes." Spinning reserves are valuable to a utility or power system because they contribute "to grid stability helping to arrest the decay of system frequency when there is a sudden loss of another resource on the system."[19]

Since cars are designed to start rapidly, they could quickly add their power to the electric grid when needed. Utilities would pay for this service if there was a guarantee that the car could deliver juice when needed, which suggests that this would be more practical for vehicle fleets or for a corporate sponsor. Individual owners would lose some control over access to their cars if they had to connect them to the grid whenever a utility demanded. And given how rarely individual car owners are likely to be called on to make their cars available for power, it is unclear that installing grid interconnect in private homes would make sense given all the safety, cost, and logistical issues that it would raise. Car manufacturers might specify in their warranty that fuel cell engines could be connected to the grid only tens of hours per year, to ensure that the vehicle's lifetime was not compromised. Also, spinning reserves are less needed at night, which is when a homeowner's car is most dependably available. But a fleet might be able to make money with this strategy. Or a corporation might sponsor an incentive for its employees and retrofit its parking lot.

It is too soon to say whether this approach will prove practical and cost-effective. We simply do not know yet whether PEM fuel cell vehicles can be reduced in cost enough that value might be extracted by using them for spinning reserves.[20] Also, this approach would probably be limited to the first several hundred thousand fuel cell vehicles. One million fuel cell vehicles, each with a 75 kW fuel cell, represents a total of 75,000 megawatts (MW), which is more than 10 percent of current total U.S. generation capacity. It is probably also more spinning reserve than the electric system will require, especially post-2020, when, it is hoped, application of advanced technology to transmission, distribution, and control of the electric grid will allow it to utilize existing generation assets more effectively. Nonetheless, this is a potentially important cash stream to help subsidize the first generation of fuel cell vehicles, which otherwise will be quite expensive to buy and operate. Some early pilots should test the feasibility of using cars in this manner.

The corollary to using fuel cell cars as mobile utilities is using a stationary fuel cell to cogenerate hydrogen and electricity or to trigenerate electricity, heat, and hydrogen. This is an option that a number of high-temperature fuel cell companies are pursuing. If high-temperature fuel cells can be made into successful commercial products, hydrogen could be separated and purified from the flue gas at a relatively low cost, just the incremental cost of the purification and separation system. This could yield overall system efficiencies of 90 percent in converting natural gas to usable energy and usable energy carriers. Solid oxide fuel cells (SOFCs) might achieve a cost of hydrogen of $2.00 per kilogram.

This price is far below that of most other means of on-site generation, and since these fuel cells could be sited at urban fueling stations or fleet fueling stations, they avoid the need for the costly hydrogen delivery infrastructure. This represents a potential market entry strategy. It also represents a productive use of natural gas fuel that would otherwise be largely wasted or poorly utilized. We are, however, several years from knowing whether fuel cells can achieve these cost targets.

Conclusion

Reducing the cost of the hydrogen infrastructure—and solving the chicken-and-egg problem—is a major hurdle in the path to a hydrogen economy. With current technology, a hydrogen infrastructure is likely to cost several hundred billions of dollars. Who will spend the money for that infrastructure before a significant number of commercial fuel cell vehicles are available? How will vehicles achieve the production levels needed to bring costs down to affordable levels if the infrastructure does not exist? There are no clear-cut answers to these questions, but what is clear is that the government will almost certainly have to intervene in the marketplace with regulations and subsidies to make it happen. The factors that will motivate government agencies to act, that will serve to accelerate or decelerate the transition to a hydrogen economy, are the subject of the next two chapters.

CHAPTER 7

Global Warming and Scenarios
for a Hydrogen Transition

T he transition to a hydrogen economy would touch every corner of our lives and almost every aspect of our economy. How fast we make the changeover will depend on several variables, not one of them easy to predict:

- How quickly can we cut the costs and improve the performance of hydrogen production, hydrogen storage, and fuel cells?
- How severe will global warming be, and how willing will governments be to take strong action to reduce the causes of that warming?
- What will be the future price and availability of both natural gas and petroleum?
- How successful will be the "competition" to hydrogen fuel cell vehicles, such as hybrid diesel-electric cars?

In this chapter and the next I examine these questions, focusing on what is fast becoming this century's gravest threat to our nation's environmental and economic well-being—the growing possibility of rapid and irreversible climate change. Global warming will, I believe, soon be the major driver of U.S. energy policy and private sector energy-related investment.

Rather than trying to predict the future, I draw from the global benchmark for strategic planning on energy, the work of the Royal Dutch/Shell Group, and examine Shell's different scenarios for the hydrogen transition. These scenarios are striking. They have convinced Shell that, globally, "the rise in human-induced carbon dioxide emissions can be halted within the next 50 years . . . without jeopardizing economic development" and that a key role will be played by hydrogen or renewable energy or both. Yet, as we'll see, hydrogen vehicles are unlikely to contribute significantly to cutting U.S. greenhouse gas emissions for the first half of this century— although hydrogen in electric power plants may well be essential. At the end of the next chapter, I offer my own scenario for the hydrogen transition.

The Accelerating Risk of Climate Change

Of all the trends that will affect energy consumption and the role of hydrogen in this century, none can match global warming for potential impact. I will discuss the subject in some depth here because of its enormous implications for the hydrogen economy and for our environmental well-being. Our understanding of the science has improved greatly in just the past few years. In the collective view of the vast majority of leading climate scientists, we are rapidly losing the opportunity to prevent dangerous and irreversible climate change. The choices we make in this decade will affect the planet for decades to come and, secondarily, will determine the shape of the hydrogen economy.

The greenhouse effect makes possible the life we know. It is caused by a blanket of heat-trapping gases naturally found in Earth's atmosphere that keeps the planet 60°F warmer than it would otherwise be—fit for humans.[1] Since the industrial revolution, however, humankind has been emitting vast quantities of greenhouse gases into the atmosphere—a planet-wide uncontrolled experiment whose risks we are only beginning to understand. Carbon dioxide, CO_2, released in the burning of fossil fuels—such as coal, oil, and natural gas—makes up 85 percent of U.S. greenhouse gas emissions.

In 1997, representatives of the developed countries agreed in Kyoto, Japan, to reduce greenhouse gas emissions to 5 percent *below* 1990 levels by 2008–2012. The European countries, Japan, and many others have ratified the agreement, whereas the United States has made it clear that it will not. Yet, from a long-term perspective, the Kyoto cuts are the most modest of starting points. As British Prime Minister Tony Blair said in February 2003, "it is clear Kyoto is not radical enough," given the scale of the climate problem. Britain's Royal Commission on Environmental Pollution concluded that "to stop further damage to the climate . . . a 60% reduction by 2050 was essential."[2]

Even though this would force a dramatic change in the way England uses energy in transportation, industry, and buildings, Blair announced that "for Britain, we will agree to the Royal Commission's target of a 60% reduction in emissions by 2050. And I am committed now to putting us on a path over the next few years towards that target."[3] What is the basis for this remarkable statement, and for the willingness of so many other nations to embrace significant reductions in CO_2 emissions?

First, scientific evidence of climate change and global warming continues to grow relentlessly. Over the past 100 years, average annual temperatures around the globe have risen by 0.6°C (1.1°F) — a faster rate of warming than at any time in the past 1,000 years.[4] The five warmest years ever measured, according to the World Meteorological Organization (WMO), were, in decreasing order, 1998, 2002, 2001, 1995, and 1997.[5] All ten of the hottest years in the global temperature record have occurred since 1987. An August 2003 analysis, the most comprehensive to date of the planet's climate history, concluded that the warming since 1980 is "unprecedented for at least roughly the past two millennia for the Northern Hemisphere." We're hotter than we've been in 2,000 years. While some global warming skeptics have argued that the planet was warmer from about AD 800 to AD 1400, that warming turns out to have been largely confined to Europe, and "any hemispheric warmth during that interval is dwarfed in magnitude by late 20th-century warmth."[6]

The evidence of this warming can be seen around the globe, in the thinning of the Arctic ice cap (which hit a record low thickness in September 2002), the increasing rate of disintegration of the Western Antarctic Ice Sheet, the retreat of glaciers around the globe, the spread of tropical diseases to more temperate climates, rising global sea levels, and on and on. According to a report by the U.S. Department of State's Bureau of Oceans and International Environmental and Scientific Affairs, "in 1998, coral reefs around the world appear to have suffered the most extensive and severe bleaching and subsequent mortality in modern record."[7] In August 2003, a major review article in the journal *Science* noted, "The link between increased greenhouse gases, climate change, and regional-scale bleaching of corals, considered dubious by many reef researchers only 10 to 20 years ago, is now incontrovertible."[8]

As the temperature increases, leading to increased evaporation from major bodies of water, the whole climate system changes. On the one hand, we have the growing intensity of regional droughts, and on the other hand, we have regions of increased precipitation and flooding. Over the past century, the United States has witnessed a statistically significant increase in a variety of extreme weather events, including hotter summers, more severe droughts, more severe floods, and more intense twenty-four-hour downpours, as analysis by the National Climatic Data Center in Asheville, North Carolina, has shown.[9]

In July 2003, the WMO catalogued a number of these extreme events: In 2003, Switzerland experienced the hottest June in "at least the past 250 years" and the United States had 562 tornadoes in May, beating the previous record of 399 in June 1992.[10] The WMO linked these events to global climate change. As the *Independent* newspaper of London put it, the WMO "signalled last night that the world's weather is going haywire," and "the unprecedented warning takes its force and significance from the fact that it is not coming from Greenpeace or Friends of the Earth, but from an impeccably respected UN organisation that is not given to hyperbole."[11] The WMO noted, "New record extreme events occur every year somewhere in the globe, but in recent years the number of such extremes have been increasing." Since that WMO report, there have been even more extreme events, including an extended heat wave in France that the government estimates caused more than 11,000 deaths between August 1 and 15, 2003.[12]

All of us have experienced the exceedingly unusual phenomenon of weather "going haywire" over the past few decades, and science tells us it is no coincidence, no mere statistical blip. It is a pattern with a message: The climate we have come to depend on is changing—and we ourselves are the cause.

Second, the scientific consensus has grown firmer over the past decade: Global warming and climate change are being caused by human action, by our increasing release of heat-trapping gases, especially CO_2. Global emissions of CO_2 have steadily risen since the dawn of the industrial age and now amount to about 24 billion tons of CO_2 released from fossil fuel combustion each year—a five-fold jump from the 1940s.[13] While emissions might be thought of as the rate of water flowing into a bathtub, atmospheric concentrations are the water level in the bathtub. Concentrations are what affect the climate. Global concentrations of CO_2 in the atmosphere have risen from a preindustrial average of about 280 parts per million by volume (ppmv) to about 315 ppmv by 1960 to more than 370 ppmv today, and they are still steadily climbing by 1.5 ppmv per year.[14] We now have greater concentrations of CO_2 in the atmosphere than at any time in the past 420,000 years, and probably more than at any time in the past 3 million years.[15]

The evidence continues to grow that the sharp rise in these atmospheric concentrations in the past several decades and the dramatic rise in global temperatures are directly related, the first causing the second. Among the most compelling recent evidence: Data from Antarctic ice cores show a close correlation between CO_2 concentration and temperature over the past 160 millennia.[16]

To assess the state of research on global warming, the United Nations established in 1988 the Intergovernmental Panel on Climate Change (IPCC), which consists of hundreds of the world's top scientists serving as lead authors, review editors, and peer reviewers. The IPCC's third assessment concluded in 2001, "Emissions of greenhouse gasses . . . due to human activities continue to alter the atmosphere in ways that are expected to affect the climate." The IPCC added that "there is new and stronger evidence that most of the warming observed over the last 50 years is attributable to human activities."[17] Some have argued that natural factors such as variations in sunshine and volcanic eruptions, which have strongly influenced temperatures in past centuries, are the source of recent temperature trends, but more recent analysis has shown that these factors can explain only about one-quarter of the warming since 1900.[18]

President George W. Bush asked the National Academy of Sciences for a report on climate change and on the conclusions of the IPCC assessments. In June 2001, the eleven-member blue-ribbon panel, which included experts previously considered skeptical about global warming, unanimously concluded: "Greenhouse gases are accumulating in Earth's atmosphere as a result of human activities, causing surface air temperatures and subsurface ocean temperatures to rise. Temperatures are, in fact, rising." The panel noted that "the IPCC's conclusion that most of the observed warming of the last 50 years is likely to have been due to the increase in greenhouse gas concentrations accurately reflects the current thinking of the scientific community on this issue. *The stated degree of confidence in the* IPCC *assessment is higher today than it was 10, or even 5 years ago.*"[19]

Commenting on the steady stream of peer-reviewed reports and articles documenting global climate change appearing in his journal and others, *Science*'s editor-in-chief, Donald Kennedy, said in 2001, "Consensus as strong as the one that has developed around this topic is rare in science."[20]

The third reason why so many developed countries (and leading companies) are willing to pursue strong action to reduce greenhouse gas emissions is the dire news: The best scientific models project that the current path of rapidly increasing emissions will lead to changes in the global climate both dramatic and unpleasant. As the National Academy of Sciences noted in 2001, "climate change simulations for the period of 1990 to 2100 based on the IPCC emissions scenarios yield a globally-averaged surface temperature increase by the end of the century of 1.4 to 5.8°C (2.5 to

10.4°F) relative to 1990."[21] These would be dramatic changes, considering that the average global temperature has risen only 5°C since the last Ice Age. Sea level could rise between 3.5 and 35 inches by 2100, with harsh implications for coastal and island ecosystems and their human communities. Tens of millions of people worldwide would see their homes destroyed. We could expect more intense storms, more severe floods, and more devastating droughts.

The IPCC climate modelers "dramatically revised upwards the top-range limit of their predictions of global warming" from their previous assessment, as Stephen Schneider explained in 2001 in the journal *Nature*, for two primary reasons: "a few predictions containing particularly large CO_2 emissions" and "smaller projected emissions of climate-cooling aerosols."[22] Previous warming estimates had assumed that we would reduce global CO_2 emissions fast enough to stabilize global concentrations of CO_2 at twice preindustrial levels (roughly 560 ppmv). Yet it is far from clear that we will act quickly enough, and thus we may have dangerously large emissions for decades, a point I will return to later in the chapter.

Equally important, scientists are gaining a deeper understanding of just how much aerosols have been shielding us from severe climate change. Sulfate aerosols, which can come from sources such as coal plants, have a cooling effect. For instance, the "leveling off" from 1940 to 1975 of the twentieth century's general warming trend is "explained largely by the cooling effects of sulfate aerosols temporarily offsetting the warming due to increasing concentration of greenhouse gases," as Tom Wigley, senior scientist with the National Center for Atmospheric Research in Boulder, Colorado, noted in July 2003.[23] This leveling off is sometimes cited today to suggest that scientists do not understand global climate trends, but the reverse is true. The temporary halt in warming was explained in the late 1980s and, in fact, is now part of scientific modeling "that is in remarkable agreement with the observations."[24]

A 2003 workshop attended by top atmospheric scientists in Berlin concluded that the shielding effect of aerosols may be far greater than previously estimated. Nobel laureate Paul Crutzen said in June 2003, "It looks like the warming today may be only about a quarter of what we would have got without aerosols." The bad news is that this conclusion suggests the planet is far more susceptible to warming than previously thought. Crutzen noted that aerosols "are giving us a false sense of security right now."[25]

Most countries are working to reduce emissions of smoke and particulates because they harm the lungs and have such clear links to cardiopulmonary disease. Significantly, greenhouse gases such as CO_2 linger in the atmosphere for 100 years or more, while aerosols typically stay only a few days, so as we properly reduce particulates to save lives today, our shield may vanish, resulting in warming that might be even greater than the IPCC's maximum forecast of 5.8°C (10.4°F). The plausible worst-case warming range may be 7°C–10°C (12.6°F–18°F).[26] If this high-end scenario comes to pass, the results will almost surely be catastrophic.

Indeed, even before this new research, Stephen Schneider's analysis in the journal *Nature* in 2001 proposed 3.5°C as the "threshold beyond which many believe substantial climate change would occur" and said that the IPCC scenarios suggest a 23 to 39 percent chance that we may exceed that threshold.[27] These are frightening odds for a global cataclysm.

A 2002 article in the journal *Science* looked at what action would be needed to avoid damage to major ecosystems and "large-scale discontinuities" in the climate system.[28] The authors, Princeton University's Michael Oppenheimer and Brown University's Brian O'Neill, argued:

- "Coral reefs are likely to undergo annual bleaching and eventually experience severe damage if the average global temperature increases more than 1 degree Celsius." Coral reefs are home to some of the ocean's richest biodiversity.
- "The potential for the Western Antarctic Ice Sheet to disintegrate is highly uncertain, but is likely to require an increase of 2 degrees Celsius or more. Complete disintegration, which could take hundreds of years or longer, would result in an increase in sea level of 13 to 20 feet. That would submerge much of the world's coastlines, including large sections of Manhattan and southern Florida."

- A shutting down of the large-scale circulation of the oceans (the so-called thermohaline ocean current) "would probably require a 3-degree Celsius increase during this century. The extent of the disruption that would result for societies and ecosystems is uncertain," but it could, for instance, result in permanent loss of the Gulf Stream, which helps warm western Europe.

Oppenheimer and O'Neill project that if we could stabilize atmospheric concentrations of CO_2 at 450 ppmv (we're at 370 ppmv), then warming this century would probably be restricted to 1.2°C–2.3°C. That would not save the coral reefs, but it might prevent a catastrophic rise in sea level or disruption of ocean currents. Unfortunately, as we will see, stabilizing at 550 ppmv is probably the best the planet can hope for, and even that will be a political and economic challenge as great as any the world has ever faced, one that requires immediate action.[29]

The scientific research revealing that warming this century may be far worse than expected should be motivation enough to act. Yet now we have to drastically rethink the sense of urgency required. Climate change can happen fast, as detailed in a 2002 study by the National Academy of Sciences:

Recent scientific evidence shows that major and widespread climate changes have occurred with startling speed. For example, roughly half the north Atlantic warming since the last ice age was achieved in only a decade, and it was accompanied by significant climatic changes across most of the globe. Similar events, including local warming as large as 16°C, occurred repeatedly during the slide into and climb out of the last ice age.[30]

Before the 1990s, scientists tended to see climate change as a slow, gradual event, linked to changes in Earth's orbit on the time scale of tens of thousands of years and changes driven by continental drift over the course of tens of millions of years. Today we know that huge temperature swings and a doubling of precipitation have occurred "in periods as short as decades to years."[31]

Most alarming is the interplay between human activity today and future climate change:

Abrupt climate changes were especially common when the climate system was being forced to change most rapidly. Thus, greenhouse warming and other human alterations of the Earth's system may increase the possibility of large, abrupt, and unwelcome regional or global climatic events.[32]

The study notes that "climate surprises are to be expected" in the future. The National Academy of Sciences warns that "denying the likelihood or downplaying the relevance of past abrupt events could be costly."[33]

So we are in the doubly dangerous situation of facing climate change that may be more extreme and may occur more quickly than we expected only a few years ago. The energy path we are currently on is thus an increasingly risky one. In a business-as-usual scenario of expanded global reliance on the combustion of fossil fuels, global annual greenhouse gas emissions would approximately triple by the end of the century, *and the atmospheric concentration of CO_2 would increase to three times preindustrial levels* (more than 800 ppmv). No wonder so many governments and businesses are willing to take strong action.

Back in 1997, Sir John Browne, chief executive officer of British Petroleum, said, "The time to consider the policy dimensions of climate change is not when the link between greenhouse gases and climate change is conclusively proven but when the possibility cannot be discounted and is taken seriously by the society of which we are part. We in BP have reached that point."[34] BP has voluntarily reduced its emissions to 10 percent below 1990 levels, has aggressively expanded its photovoltaics company, BP Solar, and has launched a major effort to produce hydrogen.

The Planners of Royal Dutch/Shell

Another major oil company committed to action on climate change is Royal Dutch/Shell. In March 2003, Shell's chair, Sir Philip Watts, said at Rice University, "We stand with those who believe there is a problem, and that it is related to the burning of fossil fuels, [and] we stand with those who are prepared to take action to solve that problem now, before it is too late."[35]

Shell is always worth listening to. As the *Economist* magazine has noted, "the only oil company to anticipate both 1973's oil-price boom and 1986's bust was Royal Dutch/Shell."[36] Throughout the 1990s, Royal Dutch/Shell found itself atop the list of the world's most profitable companies, thanks in large part to its excellence in strategic planning. In 2002, the company had sales of $179 billion, profits of more than $9 billion, and more than 100,000 employees in 132 countries.

Bear in mind that Shell's whole business centers on the extraction and delivery of fossil fuels. How remarkable, then, that the scenarios Shell has developed have convinced it that the world can respond to global warming while maintaining historical levels of economic growth. Shell's long-term energy scenarios go out to 2050. The basis for the scenarios is stabilization of CO_2 concentrations in the atmosphere at just below 550 ppmv (that is, just below a doubling of preindustrial levels). In 1995, Shell's scenarios achieved this goal in one of two ways: (1) very aggressive increases in energy efficiency or (2) very aggressive increases in renewable energy. These scenarios helped lead Shell to limit its own greenhouse gas emissions, to invest half a billion dollars in its new core renewable energy business, and to launch Shell Hydrogen. Six years later, in 2001, Shell produced very different scenarios. These are of special interest here in part because they highlight the potential role of hydrogen and renewable energy in reducing greenhouse gas emissions and in part because they reveal just how difficult it will be for the world to prevent catastrophically dangerous climate change.

Shell's 2001 Scenario One: "Dynamics as Usual"

This scenario represents "a world where social priorities for *clean, secure,* and, ultimately, *sustainable* energy shape the system."[37] Although fuel cell vehicles are introduced in 2005, "fueling inconvenience and consumer indifference limit their use to fleets." Instead, existing technologies respond to the challenge, leading to super-efficient vehicles based on hybrid electric cars and cleaner diesel engines "using as little as a third of the fuel to deliver the same quality ride" and ultimately costing less than conventional vehicles.

Use of natural gas increases steadily through 2025 and then stalls because of supply constraints. Renewable energy also soars, and "by 2020 a wide variety of renewable sources is supplying a fifth of electricity" in many developed countries. Growth stalls because of cost and logistical concerns, but by 2025 "biotechnology, materials advances and sophisticated electric network controls have enabled a new generation of renewable technologies to emerge." Advances in storage technology also facilitate a renewables renaissance. Oil becomes scarce around 2040, but super-efficient vehicles running on liquid biofuels from biomass plantations solve the problem, possibly with help from super-clean diesel fuel made from natural gas. "By 2050 renewables reach a third of world primary energy and are supplying most incremental energy."

Shell is positing dramatic increases in both renewable energy and energy efficiency for the next five decades. By way of comparison, in the United States, the "new" renewables Shell focuses on, such as wind and solar power, currently represent less than 1 percent of electricity, and the U.S. federal government has been unwilling to enact a mandate that renewables account for even 20 percent of electricity by 2020. The efficiency gains under this scenario are equally remarkable. From 1975 to 2000, world gross domestic product (GDP) slightly more than doubled and primary energy use grew by about 60 percent. From 2025 to 2050 in Shell's "Dynamics as Usual" scenario, GDP nearly doubles, but primary energy use grows by only 30 percent. This means that governments, industries, and consumers have to be embracing energy efficiency and fuel-efficient vehicles at an unprecedented rate.

This scenario shows that the path to stabilizing CO_2 emissions near 550 ppmv—which still could very well result in devastating global climate change—will be an exceedingly difficult one. From my perspective, this scenario represents not "Dynamics as Usual" but the quantum leap combination of major technological break-throughs in all areas of renewable energy, energy storage, and energy efficiency coupled with very aggressive action by governments to advance these technologies and limit greenhouse gas emissions. In other words, Scenario One is an abrupt change from the past. It might better be labeled "Unusual Dynamics."

Shell's 2001 Scenario Two: "The Spirit of the Coming Age"

This scenario "is the story of a solution that triggers a technology revolution"—the hydrogen revolution. The "solution" is a techno-logical fix for the chicken-and-egg problem: the development of "a new 'fuel in a box' for fuel-cell vehicles"—two-liter bottles capable of carrying enough fuel to drive forty miles.

"Fuel boxes can be distributed like soft drinks to multiple distri-bution channels, even dispensing machines. Consumers can get their fuel anywhere and any time." By 2025, one-quarter of the industrialized vehicle fleet uses fuel cells, which also account for half of new sales. Renewables start out slowly but pick up speed after 2025. Some one billion metric tons of CO_2 are sequestered in 2025, and, after 2025, hydrogen "is widely produced from coal, oil and gas fields, with carbon dioxide extracted and sequestered cheaply" at the source. Also, "large-scale renewable and nuclear energy schemes to produce hydrogen by electrolysis become attrac-tive by 2030."

This scenario has very little energy efficiency. Global energy use nearly triples from 2000 to 2050. Global nuclear power production also nearly triples during that time. Natural gas turns out to be wildly abundant in this scenario, and its use more than triples dur-ing those five decades. But renewable energy is also abundant. There is actually more renewable energy produced in 2050 in this scenario than in "Dynamics as Usual." By 2050, CO_2 sequestration exceeds 8 billion metric tons per year, one-fifth of emissions. That is, the world is sequestering more CO_2 in 2050 than the United States produces today from all of its coal, oil, and natural gas combustion.

If any scenario shows how difficult it will be to stabilize CO_2

emissions at 550 ppm, this is the one. Shell is quick to say that the scenarios are not predictions but thought-provoking exercises, yet this one is uncharacteristically reliant on a series of technological fixes.

The scenario says that "fuel in a box breaks distribution paradigms," but it never says what this magical box is or even what form of hydrogen it contains. It could not be any technology we are close to commercializing today, such as high-pressure storage, since that would be too bulky, too heavy, and almost certainly too dangerous to distribute by vending machine. It could not be metal hydrides, since that would be even heavier. And it could not be liquid hydrogen, since that could not be dispensed in small, portable, lightweight bottles. None of these could be easily used by the consumer to fuel a hydrogen vehicle.

The scenario says: "Fuel cell sales start with stationary applications to businesses willing to pay a premium to ensure highly reliable power without voltage fluctuations or outages. This demand helps drive fuel cell system costs below $500 per kW [by 2006], providing a platform for transport uses in stimulating further cost reductions" to $50 per kilowatt by 2010. Yet, as we have seen, the high-reliability market is unlikely to drive demand and cost reductions, especially for proton-exchange membrane (PEM) fuel cells. The scenario has many such questionable assertions and assumptions.[38]

Jeroen van der Veer, Shell's vice chairman, said in April 2003 remarks on this scenario, "We estimate that the initial investment required in the US alone to supply just 2% of cars with hydrogen by 2020 is around $20 billion."[39] This suggests that the scenario of serving 25 percent of cars in 2025 would be very costly. And, indeed, Shell Hydrogen's former chief executive officer, Donald P. H. Huberts, told the House Science Committee in March 2003, "Further build-up of the hydrogen infrastructure [beyond 2 percent] will require hundreds of billions of US dollars."[40]

Finally, according to this script, in 2025 the world is sequestering 1 billion metric tons of CO_2 per year while simultaneously producing hydrogen and shipping it hundreds of miles for use in cars. This is equivalent to sequestering the CO_2 produced by more than 700 medium-sized generation units, about two-thirds of all coal-fired plants in the United States today. Recall, however, from Chapter 4 that in February 2003 the U.S. Department of Energy (DOE) announced the billion-dollar, ten-year FutureGen project to design, build, construct, and demonstrate a 275-megawatt prototype plant that would cogenerate electricity and hydrogen and sequester 90 percent of the CO_2.[41] The goal of this effort is to "validate the engineering, economic, and environmental viability of advanced coal-based, near-zero emission technologies that by 2020" will produce electricity that is only 10 percent more expensive than current coal-generated electricity. That is, the United States hopes to have validated the viability of a breakthrough system only five years before Shell anticipates that 700 of them will be built worldwide.

You can judge how plausible this all is. From my perspective, this scenario should be called not "The Spirit of the Coming Age" but "Deus ex Machina." Every major obstacle to a hydrogen economy and to CO_2 reductions is eliminated with a magic wand. Breathtaking breakthroughs occur quickly in countless technologies—hydrogen production and storage, fuel cells, solar energy, biofuels, and sequestration. Governments willingly spend hundreds of billions of dollars to bring these technologies to the marketplace. Political obstacles to tripling nuclear power production evaporate. Natural gas supplies are seemingly limitless.

In previous books and articles, I have written enthusiastically about other Shell scenarios, and the company certainly deserves praise for its investments in renewable energy and hydrogen as well as its leadership in reducing greenhouse gas emissions. But these new scenarios seem quite flawed in ways that undermine their claims to change the way we should think about the future. The most glaring omission is any discussion of cost-effectiveness. Both of these scenarios clearly show just how much effort will be required to stabilize CO_2 concentrations at 550 ppmv, but they seem to imply that we can rely on technological luck rather than decades of tough choices. We can assume that these scenarios were designed to stimulate discussion—and they have, but in so doing the usually reliable Royal Dutch/Shell Group has not given us a productive way to look at either the problem of global warming or the solutions offered by technologies such as hydrogen and fuel cells.

In Shell's hydrogen scenario, CO_2 emissions peak in 2025 and start to decline shortly thereafter, in large part because of carbon sequestration at a massive scale. Yet this makes sense only in the context of a price for CO_2, either a tax on CO_2 emissions or a price for a permit to emit CO_2 similar to the system used to trade emissions of sulfur dioxide under the Clean Air Act administered by the U.S. Environmental Protection Agency. Sequestration would then make economic sense because it would avoid the need to pay the tax or would allow a company to sell a permit for money. If there were no value in sequestering CO_2, few would go to the expense of separating, purifying, transporting, and permanently storing it. Such carbon capture and storage costs are currently in the range of $30 to $80 per ton of CO_2, according to the DOE.[42] Significant research and development is being carried out to bring the costs down dramatically. I will discuss this issue further in the next chapter.

Reversing the global growth in CO_2 emissions by 2025, virtually all analysts would agree, will require a significant price for CO_2 as well as tough government regulations and tens, if not hundreds, of billions of dollars in subsidies that will drive businesses and consumers to save energy and choose different fuels and power sources.

Conclusion

A growing body of peer-reviewed scientific research, together with consensus reports by large groups of leading scientists, makes the case that humans are significantly warming the globe and changing the climate and that the situation is going to get much worse, with potentially catastrophic consequences for all of us. Energy technologies that reduce greenhouse gas emissions will be a critical part of the world's response, and scenarios can be a powerful tool for imagining different futures and the role technologies such as hydrogen can play.

The Shell scenarios, however, never pose a fundamental question that the nation and the world must answer as soon as possible: What are the most cost-effective strategies for reducing CO_2? We don't have an unlimited amount of money to spend, so it will be critical to maximize the environmental benefits of the money that we do have. And we are fast running out of time.

CHAPTER 8

Coping with the Global Warming Century

During my five years at the U.S. Department of Energy (DOE), the two questions I was asked most often were "How expensive will it be to reduce U.S. greenhouse gas emissions?" and "What role can clean energy technologies play in reducing that cost?" These questions became increasingly important in the months leading up to the 1997 international negotiations in Kyoto, Japan, as President Bill Clinton's administration tried to determine what level of emissions reductions it could agree to without jeopardizing the U.S. economy.

Many traditional economic analyses concluded that the cost of reducing U.S. greenhouse gas emissions to 1990 levels by 2010 would harm the economy and cost jobs, with carbon dioxide (CO_2) permits costing $60 per metric ton or more, which would raise energy prices by as much as 40 percent. These results made use of "top-down" economic models that rely on macroeconomic assumptions about how fast technology changes and are thus especially weak in their ability to characterize the effects of technology. Years earlier, most of these same models had wildly overestimated the price of industrial permits to emit sulfur dioxide that would result from the Clean Air Act restrictions, some by a factor of five or more.[1]

After I was put in charge of technology analysis at the DOE's Office of Energy Efficiency and Renewable Energy (EERE), it seemed more practical to have a "bottom-up" analysis that considered in detail the specific technologies that could reduce U.S. CO_2 emissions and their likely cost and benefit to the nation.

Expert technologists at the national laboratories were the ideal authors for such a study. The laboratories, together with EERE staff, are responsible for the vast majority of federal research and development into hydrogen and fuel cells; energy-efficient building and industrial technologies; cogeneration and distributed energy; and all forms of renewable energy, including solar, wind, and biomass. These are a large fraction of the core technologies that will be vital to reducing greenhouse gas emissions.

For a full year, I supervised an effort by some of the best analysts at five U.S. national laboratories. The result was the September 1997 report *Scenarios of U.S. Carbon Reductions: Potential Impacts of Energy-Efficient and Low-Carbon Technologies by 2010 and Beyond*. It concluded that significant emissions reductions were possible for the country with no net increase in the nation's energy bill, plus a much lower price for CO_2 emissions reductions, $7 or $14 per ton—a long way from the $60 per ton predicted by others.[2] Achieving this would take a significant effort by the government to accelerate the deployment of a variety of clean energy technologies, but the conclusion was based on existing commercial or near-commercial products—energy-efficient lighting, cogeneration, hybrid electric vehicles, wind power. It would require no major technology breakthroughs.

In November 2000, the DOE released an even more comprehensive analysis by our national laboratories extending out to 2020, which showed that U.S. CO_2 emissions could, with serious government effort, be reduced dramatically below 2000 levels, again with a price for CO_2 of less than $15 per ton and with reductions in energy bills to consumers and businesses that exceeded the economic costs.[3] Other benefits included reductions in urban air pollution and oil imports.

One conclusion I draw from all this analysis is that, in the near term, hydrogen almost certainly will not be able to compete with the myriad alternative strategies for reducing greenhouse gas emissions. Why? Four reasons:

- The competition—more fuel-efficient internal combustion engine vehicles—is getting tougher, so the incremental benefits of using fuel cell vehicles will be smaller than if they were replacing existing vehicles.
- In the near term, hydrogen is likely to be made from fossil fuel sources, which entails significant greenhouse gas emissions.
- The annual operating costs of a fuel cell vehicle are likely to be much higher than those of the competition for the foreseeable future.
- The fuels used to make hydrogen for transportation could achieve larger greenhouse gas savings at lower cost if used instead to displace the dirtiest stationary electric power plants.

In this chapter, I examine each of these points as well as other potential drivers of a transition to a hydrogen economy. I end with my own scenario for the transition.

The Competition

Underestimating the competition may be the single biggest reason why new clean energy technologies achieve success in the market much more slowly than expected. Consider renewable energy, which has been the focus of considerable public and private research and development (R&D) in the past few decades. In a 1999 study of five major renewable technologies as well as conventional power generation, Resources for the Future concluded:

In general, renewable technologies have failed to meet expectations with respect to market penetration. One exception to this trend is wind, which has met projections from the 1980s, although earlier projections were overly optimistic. The other exception is biomass applications, for which market penetration has exceeded previous projections.[4]

The study noted that these failures came in spite of the fact that R&D efforts did reduce the cost of renewables:

Renewable technologies have succeeded in meeting expectations with respect to cost. For every technology analyzed, successive generations of projections of cost have either agreed with previous projections or have declined relative to them.

Government R&D for renewables has been exceedingly successful, bringing down the cost of many renewables by a factor of ten in only two decades—in spite of the fact that the R&D budget for renewables was cut by 50 percent in the 1980s and didn't return to comparable funding levels until the mid-1990s. For example, a separate, earlier analysis in *The Technology Pork Barrel,* which examined more than a dozen major government R&D efforts, including the space shuttle and the nuclear breeder reactor, concluded, "Unlike the other technologies discussed in this book, PV [photovoltaics] had very good outcomes."[5]

Why, then, have most renewables not achieved the level of marketplace success that was projected? The authors conclude, "To a significant degree, the difference in performance in meeting projections of penetration and cost stem from the declining price of conventional generation, which constitutes a moving baseline against which renewable technologies have had to compete." The competition got tougher.

In fact, not only did traditional electricity generation reduce costs in the 1980s and 1990s (rather than increasing costs, as had been projected in the 1970s), but it did so while reducing emissions of urban air pollutants. Also, as noted in the discussion in Chapter 3 of on-site power generation, which involves many forms of renewables, utilities place many barriers in the path of new projects, and these technologies typically receive little or no monetary value for "the contributions they make to meeting power demand, reducing transmission losses, or improving environmental quality."[6]

The competition from renewables does drive the entire power generation industry to improve its performance. And renewables, especially wind power, have become a major marketplace success in

Europe, providing 20 to 40 percent of electric power in parts of Germany, Denmark, and Spain. Although renewables have been slow to achieve market share in the United States, they will very likely become essential components of the nation's and the world's efforts to avoid catastrophic global warming. As such, they represent a significant long-term success for government R&D.

The internal combustion engine, running on gasoline, has dominated the transportation market for nearly a century. Significant advances in both the engines and their fuels (such as reformulated gasoline) have dramatically reduced the urban air pollution of internal combustion engine cars, helping them fight efforts by competitors such as electric battery cars and natural gas vehicles. In the area of reduced greenhouse gas emissions, the direct competition to fuel cell vehicles includes hybrid electric vehicles and diesels, which themselves are the subject of considerable R&D today.

Hybrid gasoline-electric cars can be twice as efficient as regular cars. The onboard energy storage device, usually a battery, increases efficiency in several ways. It allows "regenerative braking"—recapture of energy that is normally lost when the car is braking. It also allows the engine to be shut down when the car is idling or decelerating. Finally, because gasoline engines have lower efficiencies at lower power, the battery allows the main engine to be run at higher power and thus more efficiently more of the time, especially in city driving.

The first-generation Toyota Prius has a city mileage of 52 miles per gallon (mpg) and a highway mileage of 45 mpg. The second-generation Prius, debuted in 2003, has even better mileage. Toyota plans to introduce a great many more hybrid models in the coming years, and most automakers will be coming out with their own hybrid vehicles. Imagine how good hybrid vehicles will be by 2020, when they might first have to face fuel cell vehicles in the marketplace.

The high efficiency that hybrids have in urban settings poses particularly tough competition for fuel cell vehicles because, at least initially, fuel cell vehicles are likely to be used mainly for urban driving, for several reasons:

- Early models probably will not have the driving range of regular vehicles.
- Early models probably will be used by fleets, which operate mainly in cities.
- The limited number of fueling stations early in their deployment will restrict long-distance travel.[7]

Fuel cells typically have higher efficiencies at lower power, so hybridizing a fuel cell vehicle (by adding a battery) does not improve its efficiency as much as does hybridizing a gasoline engine.[8]

The other major competition to fuel cell vehicles is diesels. Diesel engines are the workhorses for big trucks and construction equipment because of their high efficiency and durability. Modern diesel engines are quite different from the smoky, noisy engines of the 1970s and 1980s, with advances such as "electronic controls, high-pressure fuel injection, variable injection timing, improved combustion chamber design, and turbo-charging."[9] Although they represent less than 1 percent of car and light truck sales in the United States, diesels are becoming the car of choice in Europe, where gasoline prices are much higher, fuel taxes favor diesel use, and tailpipe emissions standards are less stringent. Diesels have some 40 percent of the market for cars in Europe, and by 2001 they represented the majority of new cars sold in a great many European countries. They are 30 to 40 percent more fuel efficient than gasoline vehicles. Also, producing and delivering diesel fuel releases 30 percent lower greenhouse gas emissions than producing and delivering gasoline with the same energy content.[10] While diesels currently have higher emissions of particulates and oxides of nitrogen, they are steadily reducing their emissions. Many believe that with the large amount of R&D funding currently aimed at diesels, they will be able to meet the same standards as gasoline engines.

Building a clean hybrid diesel-electric vehicle will require advances in technology, but ones that seem less daunting than the major breakthroughs that will be required for a practical and affordable hydrogen fuel cell vehicle, such as advances in hydrogen production and storage technology as well as in proton exchange membrane (PEM) fuel cells. It would be unwise to assume that such a hybrid car cannot be built by 2020. Trucks have less stringent emissions requirements than cars. FedEx Express and Environmental Defense teamed up in 2000 to develop a new generation of pickup and delivery trucks running on hybrid diesel-electric engines. These trucks are expected to improve fuel efficiency by 50 percent, reduce emissions of particulates by 90 percent, and reduce emissions of nitrogen oxides by 75 percent. By September 2002, two suppliers had delivered prototypes.[11]

If Hydrogen Comes from Natural Gas

In a near-term deployment scenario, hydrogen would almost certainly be produced from natural gas and probably at local fueling stations (see Chapter 5). This has three effects on greenhouse gas emissions. First, the extraction and delivery of natural gas to a fueling station entails very small leaks of natural gas, important because methane—the primary component of natural gas—has twenty-one times the global warming effect of CO_2. Second, reforming natural gas into hydrogen not only produces CO_2 but also is an inefficient process, particularly for the kind of small reformers that would be seen at filling stations. You might expect to lose as much as one-third of the energy in the natural gas.[12] Third, compression of hydrogen, the likely near-term approach for onboard storage, requires lots of electricity, the production of which also releases significant greenhouse gases. The life-cycle well-to-wheel efficiency of such a fuel cell vehicle, in terms of energy delivered to the wheels divided by total energy input, might be only 25 percent, according to one 2002 analysis, though future advances might increase that a little.[13]

In 2003, the Massachusetts Institute of Technology did a comprehensive "assessment of new propulsion technologies as potential power sources for light-duty vehicles that could be commercialized by 2020," focusing on life-cycle energy consumption and greenhouse gas emissions.[14] The assessment analyzed hydrogen generated from natural gas and concluded that a diesel hybrid would have greenhouse gas emissions more than 10 percent *below* those of a hydrogen fuel cell vehicle and would have roughly the same emissions as a hybridized fuel cell vehicle. Equally noteworthy, *the projected gasoline hybrid for 2020 has roughly the same life-cycle greenhouse gas emissions as the hydrogen fuel cell vehicle.* This is one of many reasons arguing against significant investment in natural gas reforming and local fueling stations.

Annual Operating Costs

Cost savings—economic payback—appears to be the single biggest determinant of success for a new energy technology. Renewable energy has many superior attributes similar to those of fuel cells, including zero emission of urban air pollution, but renewables have been slow to catch on in the United States because of their high cost. On the other hand, the other major area of R&D conducted by DOE's office of Energy Efficiency and Renewable Energy—developing energy-efficient technologies that reduce energy bills—has been wildly successful.

The national laboratories, funded by EERE, developed a host of more efficient devices, including a more efficient refrigerator, improvements to the compact fluorescent light bulb, solid-state lighting controls for office fluorescent lights, and super-efficient windows. Many of these products have achieved very significant market penetration. The National Academy of Sciences examined dozens of case studies of EERE-funded technologies and found that they had cumulatively saved American businesses and consumers $30 billion in reduced energy bills.

The products that were most successful not only reduced pollution from electricity generation but also had a good economic payback combined with equivalent or superior attributes to the products they replaced. For instance, solid-state lighting helps businesses cut lighting energy use in half or more while delivering higher-quality light without the annoying flicker of earlier fluorescents. Likewise, greenhouse gas emissions from lighting are cut in half, often with paybacks of less than two years.

Unfortunately, for the foreseeable future, hydrogen cars will almost certainly be unable to compete with alternative strategies for reducing greenhouse gas emissions. As we saw in Chapter 6, we do not even know today whether a practical and affordable fuel cell vehicle can be built. A 2002 analysis for the DOE concluded that even with optimistic assumptions about technology improvements, future fuel cell vehicles will probably cost 40 to 60 percent more than conventional vehicles.[15]

Also, hydrogen will be much more expensive than gasoline. As previously discussed, hydrogen provided at fueling stations would probably cost $4 or more per kilogram (kg). The equivalent-energy price of gasoline hovers around $1.50–$2.00, including taxes, in the United States today (since a kilogram of hydrogen has about the same energy content as a gallon of gasoline). Ultimately, if hydrogen were to be the main transportation fuel, it would itself have to be taxed unless we found a new source for funding road projects. So, even with a more efficient engine, the annual fuel costs are likely to be considerably higher, perhaps by a factor of two.

Moreover, compared with the competition, hybrids and diesels, the cost differential is even more. While hybrid and clean diesel vehicles may cost more than current internal combustion engine vehicles, at least when they are first introduced, their greater fuel efficiency means that, unlike hydrogen fuel cell vehicles, they may pay for that extra up-front cost over the lifetime of the vehicle. This means that hybrids and diesels will very likely have roughly the same annual operating costs as current internal combustion engine vehicles, a significant marketplace advantage over fuel cell vehicles.

Hence, from a policy perspective, hybrids and diesels would reduce transportation CO_2 emissions at a far lower cost per ton. The average new car today generates about four to five metric tons of CO_2 per year. Perhaps the key reason for replacing gasoline engines is to lower that number. *A fuel cell vehicle in 2020 might reduce CO_2 emissions at a price of more than $200 per metric ton (regardless of what fuel the hydrogen was produced from).*[16] *An advanced gasoline engine could reduce CO_2 for far less and possibly for a net savings because of the reduced gasoline costs.* Energy efficiency strategies in other sectors are similarly low in cost. A June 2003 analysis in the journal *Science* by David Keith of Carnegie Mellon University and Alexander Farrell of the University of California, Berkeley, put the cost of CO_2 avoided by fuel cells running on zero-carbon hydrogen at more than $250 per ton, even with optimistic reductions in fuel cell costs.[17]

Moreover, we still have the question of who will pay for the hydrogen infrastructure, which in the early going could easily come to $5,000 per car.[18] Those who believe, as I do, that global warming is the most intractable and potentially catastrophic environmental problem facing the nation and the planet this century—and therefore the problem that requires the most urgent action on the part of government and the private sector—may conclude that spending this kind of money on building a hydrogen infrastructure could take away a massive amount of resources from far more cost-effective measures. Ultimately, a hydrogen infrastructure may be critical to helping us achieve the kind of deep CO_2 reduction that we will need in the second half of this century, but probably not before then.

Better Uses for the Fuels

In the first half of the twenty-first century, the fuels used to make hydrogen for transportation could achieve far larger greenhouse gas savings at lower cost, displacing emissions in electricity generation, as several analyses have shown.[19] This is true not only for natural gas but also for renewable power in the foreseeable future.

Let's start with natural gas. As we've already seen, a fuel cell vehicle running on hydrogen produced from natural gas will have little or no net greenhouse benefits compared with the likely competition in 2020, hybrid vehicles. Natural gas, however, has a huge benefit when used to avoid the need to build new coal plants—or to displace existing ones. Coal plants do not merely have much lower efficiencies than natural gas plants—30 percent versus 55 percent. Compared with natural gas containing the same amount of energy, coal has nearly double the CO_2 emissions, whereas gasoline has only about one-third more CO_2 emissions than natural gas. A megawatt-hour (MWh) of electricity from a combined cycle gas turbine releases about 810 pounds (370 kg) of CO_2, whereas the same megawatt-hour of even relatively new coal plants can release more than 2,200 pounds (1,000 kg) of CO_2. And the savings would be even larger if the natural gas were used in a future stationary fuel cell and gas turbine system that might have 70 percent efficiency. If the nation had an unlimited supply of cheap natural gas, we could use it for all purposes, but we do not.

The United States has recently been sprinting to build new natural gas power plants because they are so efficient and clean. As of 2003, the country had more than 800 gigawatts (GW) of central station electric power generation. A gigawatt is equal to 1,000 megawatts (MW) and is the size of one very large existing power plant or three typically sized new power plants. "Of the 144 gigawatts added between 1999 and 2002, 138 gigawatts is natural-gas-fired, including 72 gigawatts of efficient combined-cycle capacity and 66 gigawatts of combustion turbine capacity, which is used mainly when demand for electricity is high," as noted by the U.S. Department of Energy's Energy Information Administration (EIA) in its *Annual Energy Outlook 2003*. As rising demand leads to rising prices for natural gas, however, the EIA forecasts an increase in coal-powered production. The EIA projects that between 2001 and 2025, 74 GW of new coal-fired capacity will come online. Significantly, "as natural gas prices rise later in the forecast, new coal-fired capacity is projected to become more competitive, and 91 percent of the projected additions of new coal-fired capacity are expected to be brought on line from 2010 to 2025."[20]

From a global warming perspective, the right approach would be to defer that increased coal-fired capacity and to spur the retirement of existing coal plants. Yet not only is coal-fired capacity projected to grow, but also the EIA projects that existing coal plants will be used far more. From 2001 to 2025, the EIA projects a remarkable 40 percent increase in coal consumption for electricity generation, which by itself would increase U.S. greenhouse gas emissions by 10 percent. Dale E. Heydlauff, senior vice president for governmental and environmental affairs at American Electric Power (AEP), a leading U.S. producer of coal-fired power, said in August

2003, "The amount of AEP's coal plant retirements will be driven primarily by the price of natural gas."[21]

Unfortunately, by 2003, rising demand for natural gas had already hit North American supply constraints, driving up prices. In June 2003, Alan Greenspan, chairman of the Federal Reserve Board, took the unusual step of testifying before the House Energy and Commerce Committee about natural gas, warning, "We are not apt to return to earlier periods of relative abundance and low prices anytime soon."[22] While robust government efforts to push more efficient use of natural gas and to accelerate renewable power are the policies that I would recommend (see the Conclusion), such approaches lack political support, and Greenspan himself did not propose them. Noting that "Canada, our major source of imported natural gas, . . . has little capacity to significantly expand its exports," he endorsed an expansion of the capacity to import liquefied natural gas (LNG).

While not as energy-intensive a process as liquefying hydrogen, cooling natural gas to a temperature of about −260°F and transporting the resulting liquid has "an energy penalty of up to 15%," according to the Australian Greenhouse Office.[23] So if LNG represents a significant fraction of incremental U.S. natural gas consumption in the future, the energy lost in the process of making and shipping LNG provides one more reason why we should not make hydrogen from natural gas, which already offers little or no energy or greenhouse gas benefit over hybrid vehicles.[24] And since we should start thinking of the energy resource base as a global one, it would be far better to use foreign natural gas to offset foreign coal combustion than to import it into the United States in order to turn it into hydrogen to offset domestic gasoline consumption. That's especially true because projected growth in global coal consumption is an even bigger greenhouse gas problem than projected growth in U.S. coal consumption.

By 1999, the world had just more than 1,000 GW of coal-fired electricity generating capacity, of which about one-third was in the United States. Between 2000 and 2030, more than 1,400 GW of new coal capacity will be built, according to the International

Energy Agency, of which 400 GW will represent replacement of old plants (see Figure 8.1).

These plants would commit the planet to total CO_2 emissions of some 500 billion metric tons over their lifetime, unless "they are backfit with carbon capture equipment at some time during their life," as David Hawkins, director of the Natural Resources Defense Council's Climate Center, told the U.S. House Committee on Energy and Commerce in June 2003.[25] Hawkins continued:

> To put this number in context, it amounts to half the estimated total cumulative carbon emissions from all fossil fuel use globally over the past 250 years! If we build any significant fraction of this new capacity in a manner that does not enable capture of its CO_2 emissions we will be creating a "carbon shadow" that will darken the lives of those who follow us.[26]

The carbon shadow is not merely the long lifetime of coal plants (many decades) but also the long lifetime of heat-trapping CO_2 in the atmosphere (more than a century).

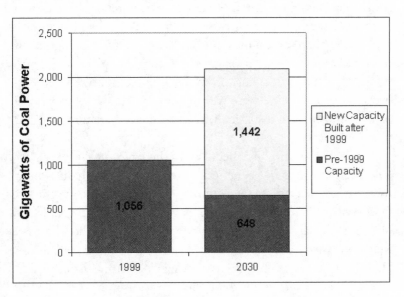

FIGURE 8.1. Two-thirds of world coal capacity in 2030 is not yet built.
Source: IEA, NRDC.

Carbon capture and storage (CCS) is an important focus of research, but on a massive scale it is unlikely to be practical and economical for more than two decades (as discussed later in this chapter). To defer as many of these plants as possible until CCS is ready, we will have to use our electricity more efficiently (thereby slowing the demand for such power plants) and build as many cleaner power plants as possible. Again, natural gas is far more critically needed for this task than for generating hydrogen. So, too, with renewable power—for the foreseeable future, we will need it in order to avoid building coal-fired plants, not to generate hydrogen for transportation.

As discussed in Chapter 4, converting electricity from renewable sources to hydrogen through electrolysis is a relatively inefficient process, immediately costing some 30 percent of the energy in the electricity. Transporting and storing the hydrogen costs another 10 to 20 percent of that energy. This wasted energy would be better used to displace fossil fuels, according to a November 2002 study by three leading analytical groups in the United Kingdom.[27] That study calculated that 1 MWh of electricity from renewables, if used to manufacture hydrogen for use in a fuel cell vehicle, would save slightly less than 500 pounds of CO_2. That is some 300 pounds *less* than the savings from displacing a future gas plant and 1,700 pounds less than the savings from displacing coal power (see Figure 8.2). And, as we've seen, we are going to build a lot more coal plants unless we can displace them with efficiency, natural gas, and renewables.[28]

The British study concluded, "Until there is a surplus of renewable electricity, it is not beneficial in terms of carbon reduction to use renewable electricity to produce hydrogen—for use in vehicles, or elsewhere. Higher carbon savings will be achieved through displacing electricity from fossil fuel power stations."[29] The July 2003 Keith and Farrell analysis in *Science* comes to a similar conclusion: "Until CO_2 emissions from electricity generation are virtually eliminated, it will be far more cost-effective to use new CO_2-neutral electricity (e.g., wind or nuclear) to reduce emissions by substituting for fossil-generated electricity."[30]

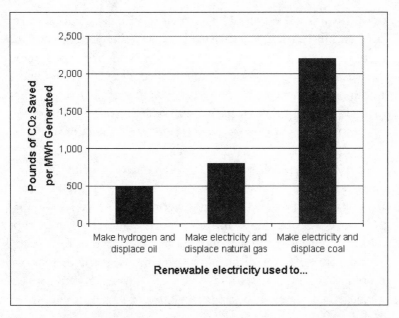

FIGURE 8.2. Emissions reduced by renewable electricity. The bar on the left represents the CO_2 savings from renewable electricity used to make hydrogen, assuming the hydrogen is used in a fuel cell car and displaces the fuel from a hybrid car. The middle bar represents the savings from renewable power displacing electricity from a combined cycle natural gas power plant. The bar on the right represents the savings from renewable power displacing electricity from a typical coal plant.

When will we have a surplus of renewable power and a virtual elimination of CO_2 emissions from electricity generation? The British authors conclude that, in the United Kingdom, it is likely to be "at least 30 years." Significantly, the United Kingdom's electric grid already has a CO_2 intensity (CO_2 emitted per megawatt-hour of electricity produced) that is lower than that of the United States by more than one-third. Moreover, the United Kingdom has moved sharply away from coal generation in the past two decades and is aggressively pursuing renewable energy and cogeneration. The U.S. government, in contrast, has been unable as of 2003 to pass a law requiring that 20 percent of electricity come from renewable power in 2020, even though the environmental benefits are very large and the economic costs small. In fact, a key reason why the costs are so small is that such a law would reduce the pressure on the natural gas supply, which would reduce prices by just about the amount of the extra electricity costs.[31] A May 2003 analysis in *Wind-power Monthly* argues that having excess U.S. wind power generation to use for hydrogen production is "a situation not likely before 2050 at the earliest."[32]

As but one example of how far the United States and the world have to go on clean energy, consider the results of a March 2003 analysis by scientists at Lawrence Livermore National Laboratory, the University of Illinois, and New York University.[33] They concluded that if the world is to stop global temperatures from rising by more than 2°C as a result of our greenhouse gas emissions, we should be building between 400 and 1,300 MW of zero-carbon electricity generation capacity *per day* for fifty years. Yet current projections for the next thirty years are that we will build just 80 MW per day.[34] And 2°C would still very likely have a devastating effect on the planet, as we have seen. Strikingly, "if climate sensitivity is in the middle of the IPCC range," the analysis also concludes, "even stabilization at 4°C warming would require installation of 410 megawatts of carbon emissions-free energy capacity each day."[35]

I think it is possible that the United States could have virtually carbon-free electricity before 2050—with hydrogen playing a key role, as I will discuss in my scenario at the end of this chapter—but it will require both major technological breakthroughs and a sea change in U.S. government policy on global warming. First, let's examine some of the other potential drivers of the hydrogen transition.

Hydrogen and Urban Air Pollution

Another driver of a shift to hydrogen is concern about urban air pollution. The transportation sector remains one of the largest sources of such pollution—especially (1) the oxides of nitrogen (NO_X) that are a precursor to ozone smog and (2) particulates—

especially those less than ten microns in size, which do so much damage to our hearts and lungs.

In the United States, vehicle emissions (other than greenhouse gas emissions), however, have been declining steadily. Toxic emissions are being reduced through the combination of ever stricter federal and state regulations coupled with the steady turnover of the vehicle fleet. The oldest and dirtiest vehicles ultimately go out of service and are replaced by the newest and cleanest vehicles. The federal Clean Air Act Amendments of 1990 set in motion a two-pronged process. In the 1990s, Tier 1 standards dramatically reduced tailpipe emissions of new light-duty vehicles (such as cars and most sport utility vehicles, or SUVs). By 2010, Tier 2 standards will even further reduce vehicle emissions and will extend the regulations to larger SUVs and passenger vans while mandating the use of gasoline with lower sulfur content (which not only directly reduces emissions but also makes it easier for automakers to design cars that achieve further reductions). Due to its infamously smogged-in cities, California has even tougher standards for automobiles than does the United States as a whole; these standards will also be phased in during this decade.

When all these standards are in place, new U.S. cars will be exceedingly clean, at least from the perspective of emissions of *urban* air pollutants (the current Clean Air Act standards do not address vehicle CO_2 emissions). Many new cars, so-called partial zero-emissions-vehicles, will actually have tailpipe emissions cleaner than Los Angeles air. Hydrogen fuel cell vehicles, though, have no emissions besides water and thus are the ultimate in a clean car. On the other hand, they will be more costly in the beginning and will require a significant investment in infrastructure. Here is the question for any such breakthrough clean technology: Are the benefits received proportional to the money spent? Put another way, can emissions reductions beyond Tier 2 standards be achieved for less money using other strategies? In the short run, the answer is yes.

For instance, remote sensing of vehicles at a major intersection in Denver, Colorado, showed that "the worst polluting 10% of cars and trucks emit 65% of the carbon monoxide pollution, while the 50% of clean, new cars emit less than 6%."[36] Those dirty vehicles are primarily "older, poorly maintained automobiles." This evidence strongly suggests that until we make a serious effort to either repair or scrap these gross polluters, we won't affect urban air pollution much by focusing on further emissions reductions in new vehicles. Trying to lower that already low 6 percent just won't have much effect. The 2003 analysis by Keith and Farrell estimates that fuel cell vehicles would reduce oxides of nitrogen at a cost 100 to 500 times greater than that of current strategies.[37] Also, in terms of overall emissions of urban air pollutants, for the next two decades at least, we can achieve pollution reductions far more cost-effectively by cleaning up power plants and large off-road vehicles, such as construction equipment, than by investing in fuel cell vehicles. The George W. Bush administration has proposed new, tougher emissions regulations for off-road vehicles.

The long-term answer to the cost-effectiveness question will require a far better understanding of what is now unknown—the cost of super-clean vehicles, their hydrogen fuel, and the necessary infrastructure. Research, development, and demonstration of hydrogen fuel cell vehicles remain the only ways to find the ultimate answers.

Hydrogen and Energy Independence

In a February 2003 speech on hydrogen, President Bush said, "It jeopardizes our national security to be dependent on sources of energy from countries that don't care for America. . . . It's also a matter of economic security, to be dependent on energy from volatile regions of the world."[38] You might think that after fighting two wars in the Persian Gulf since 1990, after enduring a major terrorist attack funded in large part by Persian Gulf oil money, and with imports accounting for more than half of our oil consumption and our sending $100 billion per year offshore, that our first call to action would be to significantly reduce those imports. And yet we have done essentially nothing.

The two most effective strategies for permanently reducing dependence on imports—raising gasoline taxes and raising fuel efficiency standards—simply lack political support, even though they both have the added benefit that they would help address global warming and other environmental problems. The gasoline tax strategy is the one Europe has used to help constrain fuel consumption. The United Kingdom and countries such as France and Germany have gasoline taxes of more than $2.00 per gallon, more than five times the gasoline tax in the United States.[39] Such tax hikes are essentially inconceivable in this country for the foreseeable future. For instance, by the end of the Clinton administration, it was not possible to suggest increasing gasoline taxes by even a few cents—tax increases of all kinds were considered political suicide.

The fuel efficiency approach is the one this country used so successfully in the late 1970s and early 1980s, when we doubled the fuel efficiency of our fleet while making our cars safer, mandating that new cars have a fuel efficiency of 27.5 mpg. In a 2002 report to President Bush, the National Academy of Sciences concluded that automobile fuel economy could be increased by 12 percent for small cars and up to 42 percent for large SUVs with technologies that would pay for themselves in fuel savings.[40] That study did not even consider the greater use of diesels and hybrids. Studies undertaken by the national laboratories for the DOE, by the Massachusetts Institute of Technology, and by the Pew Center on Global Climate Change have concluded that even greater savings could be cost-effective while maintaining or improving passenger safety.[41] The Europeans have a voluntary agreement with automakers that will reduce CO_2 emitted per mile by 25 percent from 1996 to 2008 for the average light-duty vehicle, which equates to a vehicle fuel efficiency of almost 40 mpg. Japan has a mandatory target with similar goals.[42] Yet most observers consider such new fuel efficiency standards in this country politically infeasible for the foreseeable future: The auto companies have actively lobbied against any such standards, the White House is opposed to them, and there is little evidence of broad political support in Congress or elsewhere.

Because of this inaction, as of 2002 the fuel economy of the average vehicle on American roads was at its lowest level in two decades and likely to get worse.[43] The fuel economy laws have a loophole allowing SUVs and light trucks to average 20.7 mpg, 25 percent lower than the standard for new cars. This has allowed overall vehicle efficiency to drop as the SUV share of new vehicle sales has grown. Ford, for instance, has backed off a voluntary commitment to increase SUV fuel efficiency, and, in fact, its 2003 model year SUVs will be less fuel efficient than those of the previous year.[44] Moreover, companies known for fuel efficiency, such as Honda and Toyota, both of which have introduced hybrid vehicles into the U.S. market, have also introduced gas-guzzling SUVs.

From my years of working to influence policy on this issue, both in and out of government, the conclusion I draw from all this is that "energy independence" is a phrase to which politicians and policy makers give lip service but which is exceedingly unlikely to speed up the transition to a hydrogen economy. If the actions of Saddam Hussein and Osama bin Laden and record levels of oil imports couldn't induce lawmakers, automakers, and the general public to embrace *existing* vehicle energy efficiency technologies that will actually pay for themselves in fuel savings, I cannot imagine what fearful events must happen before the nation will be motivated to embrace hydrogen fuel cell vehicles, which will cost much more to buy, cost much more to fuel, and require massive government subsidies to pay for the infrastructure.

Ultimately, I believe, it is not fear about the growing level of oil imports that will bring about change but the inevitable exhaustion of the world's finite oil resources. Worldwide population growth, economic growth, and urbanization will dramatically increase global oil consumption in the coming decades, especially in the developing world. As farmworkers move to urban centers, much more oil will be needed. The key elements of urbanization are commuting, transport of raw materials, and construction of buildings and other infrastructure. All consume huge amounts of oil. At the same time, fewer farmers will have to feed more people, so the use of mechanization, fertilizer, and transportation will increase, consuming even more oil. In spite of all this growth in the developing world, no country will overtake the United States in oil consumption anytime soon. The EIA projects that between 2000 and 2025, we will increase our oil demand by nearly 50 percent.[45]

This acceleration in global oil demand will eventually bump up against the reality of oil as a finite, nonrenewable resource, causing global production to peak and start declining, much as production in the continental United States did decades ago. Some believe this will occur by 2010—or sooner. In a 2003 speech, Matthew Simmons, energy investment banker and Bush administration advisor, said that his analysis was leading him more and more to "the worry that peaking is at hand; not years away. If it turns out I'm wrong, then I'm wrong. But if I'm right, the unforeseen consequences are devastating. But unfortunately the world has no Plan B if I'm right."[46] In his 2001 book *Hubbert's Peak: The Impending World Oil Shortage,* Princeton University geophysicist Kenneth Deffeyes states bluntly, "There is nothing plausible that could postpone the peak until 2009. Get used to it."[47]

The Royal Dutch/Shell Group, probably the most successful predictor in the global oil business, adds a few years to those gloomy forecasts. According to Shell, "a scarcity of oil supplies—including unconventional sources and natural gas liquids—is very unlikely before 2025. This could be extended to 2040 by adopting known measures to increase vehicle efficiency and focusing oil demand on this sector."[48] Whether we will adopt these known measures or not remains to be seen. Moreover, producing conventional liquid fuels from either natural gas or unconventional sources of oil (such as Canadian oil sands) is relatively energy-intensive, and relying on these sources will significantly increase greenhouse gas emissions.

I have found that this issue—the possibility that we are nearing a production peak—gets even less traction in Washington, D.C., as a driver of energy policy than the call for energy independence. The argument is inevitably dismissed with a wave of the hand: "We've heard that for decades." In 1996, at my first congressional hearing representing the DOE, a researcher from the Massachusetts Institute of Technology testified that "you really can't see a peak, because the peak keeps moving out further and further into the future. And, people who do this kind of work are always sort of explaining that the previous peak was wrong, but now they have a new peak and it's the real peak. So, I wouldn't give too much credence to that."[49]

What would happen if we started running out of oil? Prices would start spiking, possibly with destabilizing effects on the economy. On the one hand, Deffeyes and others argue that after the peak, production may well decline at a relatively rapid rate even as demand increases, so delaying action until that moment may risk significant economic damage. On the other hand, most European countries have gasoline prices that are double ours already, with no obvious harm to their standard of living. In either event, such a peak would eventually force us to switch to much more efficient vehicles and alternative fuels. So the possibility of a production peak in the near future is an argument for being prepared with alternatives to oil through an aggressive R&D effort. But again, if we were seriously concerned about this issue, we would take up Shell's suggestion and aggressively adopt "known measures" for vehicle efficiency, such as hybrids, diesels, and biofuels, discussed later in this chapter.

Hydrogen and Sequestration

Carbon sequestration—permanent carbon storage—is a potentially crucial strategy for reducing net atmospheric CO_2 emissions and one that could dramatically accelerate introduction of hydrogen into the economy. CO_2 can be removed from the atmosphere and stored *biologically* in trees and other biomass, a strategy I will return to shortly. Sequestration can also involve removing CO_2 from power plants and storing it within the planet's *physical* systems, which, as discussed earlier, is sometimes called carbon capture and storage (CCS). The CO_2 can be captured either before or after combustion.

Geologic sequestration is the storage of CO_2 in vast, sealed underground places. Costs for geologic sequestration are currently quite high, more than $30 per ton of CO_2, according to the DOE.[50] The technical challenges of reducing those costs are significant. The report of a February 2003 workshop on carbon management by the National Academy of Sciences concluded, "At the present time, technology exists for the separation of CO_2 and hydrogen, but the capital and operating costs are very high, particularly when existing technology is considered for fossil fuel combustion or gasification streams."[51] Significant R&D is being carried out to bring the costs down.

Capturing CO_2 from a gasification stream is potentially the most intriguing option for a hydrogen economy, especially for a world awash in coal but needing to restrain greenhouse gas emissions. As discussed in Chapter 4, coal can be gasified and the resulting syngas (synthesis gas) can then be chemically processed to generate a hydrogen-rich gas for fuel and a stream of CO_2 that can be piped to a sequestration site. This hydrogen-rich gas can be combusted directly in a combined cycle power plant, or it can power a high-temperature fuel cell.

Some hydrogen can also be purified and shipped out for other uses, such as transportation. From a global warming perspective, though, the same analysis that applied to renewable energy applies here: Until the U.S. electric grid is virtually CO_2-free, this hydrogen-rich gas is far better used to displace electric power derived from fossil fuels than to be converted to very pure hydrogen, shipped hundreds of miles, and used as a transportation fuel. And, in fact, the DOE's FutureGen project for coal gasification and CO_2 sequestration envisions using the hydrogen to generate clean power, at least initially, perhaps using a solid oxide fuel cell (SOFC).[52] Coal gasification systems with CO_2 capture could achieve efficiencies of 60 percent or more using various combinations of turbines and SOFCs.[53]

The key question is where to put the CO_2. The largest potential physical reservoir is the deep oceans. However, ocean sequestration poses serious environmental risks and is unlikely to be a viable climate mitigation strategy. Tens of millions of tons of CO_2 are already injected into oil fields to enhance recovery. This strategy can be expanded for early, low-cost sequestration efforts, but oil fields are limited in size and location, and transporting CO_2 over long distances will be costly. To meet a potential demand to sequester several billion metric tons of CO_2 per year, especially in places that do not have conventional storage formations, research is focusing on finding much larger reservoirs.

Recent attention has focused on pumping highly compressed liquid CO_2, so-called supercritical CO_2, into geologic formations such as deep underground aquifers. As the report of the National Academy workshop noted, "less dense than water, CO_2 will float under the top seal atop the water in an aquifer and could migrate upward if the top seal is not completely impermeable."[54]

The problem here is that even very tiny leakage rates can undermine the environmental value of such sequestration. If we are trying to stabilize CO_2 concentrations at twice preindustrial levels, a 1 percent leakage rate could add $850 billion *per year* to overall costs by 2095, according to an analysis by Pacific Northwest National Laboratory. That study concluded, "Leakage of CO_2 from engineered CO_2 disposal practices on the order of 1% or less per year are likely intolerable as they represent an unacceptably costly financial burden that is moved from present generations to future generations." If we cannot be certain that leakage rates are below 1 percent, "the private sector will find it increasingly difficult to convince regulators that CO_2 injected into geological formations should be accorded the same accounting as CO_2 that is avoided"—avoided, that is, directly through technologies such as wind power. The study notes that, "there is no solid experimental evidence or theoretical framework" for determining likely leakage rates from different geologic formations.[55]

How long will it take before CCS emerges as a major solution to global warming? That remains uncertain. As Princeton's Bob Williams wrote in 2003, "one cannot yet say with high confidence that the CO_2 storage option is viable."[56] The technology itself is very challenging, and just as commercializing fuel cells has taken much longer and has proven far more difficult than was expected, so, too, may building large commercial coal gasification combined cycle units.[57] The DOE is aiming to "validate the engineering, economic, and environmental viability" of a system by 2020.[58]

Ultimately, this integrated gasification combined cycle (IGCC) technology will require leadership by the private sector. Yet, as the National Coal Council reported to the secretary of energy in its May 2003 report titled *Coal Related Greenhouse Gas Management Issues,* "vendors currently do not have an adequate economic incentive to invest R&D dollars in IGCC advancement" because "IGCC may only become broadly competitive with" current coal and natural gas power plants "under a CO_2-restricted scenario." Thus, "power companies are not likely to pay the premium to install today's IGCC designs in the absence of clear regulatory direction on the CO_2 issue."[59] Absent near-term restrictions on CO_2 emissions, this option will be pushed much further out to the future.

As for storing CO_2 in underground aquifers, much testing will have to be done before this approach can be considered for widespread use. Each potential major site will very likely need to be the subject of intensive long-term monitoring to guarantee it can permanently store CO_2. Very sensitive and low-cost in situ assessment and monitoring techniques must be developed to provide confidence that leakage rates are exceedingly low.

Analysis suggests that CCS could eventually eliminate a large fraction of emissions from the U.S. electricity sector for $20 to $40 per ton of CO_2, whereas using the carbon-free hydrogen as transportation fuel to cut emissions in cars could cost more than $200 per ton.[60] Ultimately, leak-proof CCS may be an essential strategy for combating global warming, one that will accelerate the transition to hydrogen economy. But, at this point, it seems unlikely to be widely used until after 2030—after the world has built an additional 1,000 GW of coal plant capacity and dramatically enhanced the prospect of catastrophic global warming.

In the near term, biological sequestration can, with appropriate accounting rules and environmental criteria, play a role in reducing atmospheric concentrations of CO_2. Probably the best-known form of sequestration is the planting of trees (or other plants) to remove CO_2 from the atmosphere and store it inside that biomass. This biological sequestration is temporary because when the plant or tree dies, it decays, returning the CO_2 to the atmosphere. Even massive planet-wide reforestation would have a limited effect on CO_2 concentrations.[61]

A biological sequestration strategy with enormous potential is to plant trees and crops for the purpose of harvesting them as renewable energy, either electric power or liquid fuel.[62] If the planting and harvesting are done in a sustainable fashion, with relatively low energy inputs, the net CO_2 released will be close to zero. Biomass can be gasified and converted to hydrogen (and electricity) in a process very similar to coal gasification. A number of biomass gasification processes are in the process of being demonstrated, although a practical commercial system remains technologically challenging. If carbon capture and storage is practical, then one could also imagine extracting the CO_2 stream from the biomass gasification process. This would turn biomass into a potent net reducer of CO_2—extracting CO_2 from the air via photosynthesis while growing and then injecting that CO_2 into underground reservoirs through CCS. This may be a critical option for the United States and the world, given the increasing risk of rapid and catastrophic climate change.[63]

Biomass can also be used to make a zero-carbon transportation fuel, such as ethanol, which is now used as a gasoline blend. Today, the major biofuel is ethanol made from corn, which yields only about 25 percent more energy than was consumed to grow the corn and make the ethanol, according to some estimates. Considerable R&D is being focused on producing ethanol from sources other than corn. This so-called cellulosic ethanol can be made from agricultural and forest waste as well as dedicated energy crops, such as switchgrass or fast-growing hybrid poplar trees, which can be grown and harvested with minimal energy consumption so that overall emissions are near zero.[64] All cars today can use a mixture of 10 percent ethanol and 90 percent gasoline, known as E10. Some 4 million flexible fuel vehicles, which can run on either gasoline or a blend with 85 percent ethanol, E85, are on the road today, but few use E85 because of its high price.

If hybrids and diesels are the toughest competition for transportation fuel cells, then cellulosic ethanol is probably the toughest competition for hydrogen in the race to develop a transportation fuel with low or no greenhouse gas emissions. The big advantage ethanol has over hydrogen is that it is a liquid fuel and thus is much more compatible with our existing fueling system. Existing oil pipelines, however, are not compatible with ethanol, so significant infrastructure spending would still be required if ethanol were to become the major transportation fuel.[65] As with hydrogen, production of ethanol will require major technological advances before it can come close to the price of gasoline on an equivalent energy basis.[66]

Probably the biggest drawback to biofuels, and to biomass energy in general, is that biomass is not very efficient at storing solar energy. Therefore, large land areas are needed to provide enough energy crops if biofuels are to provide a significant share of transportation energy. One 2001 analysis by ethanol advocates concluded that to provide enough ethanol to replace the gasoline used in the light-duty fleet, "it would be necessary to process the biomass growing on 300 million to 500 million acres, which is in the neighborhood of one-fourth of the 1.8 billion acre land area of the lower 48 states" and is roughly equal to the amount of all U.S. cropland in production today.[67] That amount of displaced gasoline represents about 60 percent of all U.S. transportation-related CO_2 emissions today, but less than 40 percent of what is projected for 2025 under a business-as-usual scenario. If ethanol is to represent a major transportation fuel in the coming decades, then U.S. vehicles will need to become much more fuel efficient. Moreover, given the acreage needed, using it for these purposes would obviously have dramatic environmental, political, and economic implications. As of 2003, there were no commercial cellulosic biomass-to-ethanol plants.

Hence, ethanol, like hydrogen, is no near-term panacea. In the long term, however, biomass-to-energy production could be exceedingly efficient with bio-refineries that produce multiple products. Lee Lynd, professor of engineering at Dartmouth College, described one such future bio-refinery where cellulosic ethanol undergoes a chemical pretreatment and then fermentation converts the carbohydrate content into ethanol as CO_2 bubbles off.[68] The residue is mostly lignin, a polymer found in the cell walls of plants. Water is removed, and the biomass residue is then gasified to generate electricity or to produce a stream of hydrogen and CO_2. The overall efficiency of converting the energy content of the original biomass into useful fuel and electricity would be 70 percent, even after accounting for the energy needed to grow and harvest the biomass. The CO_2 can be sequestered. Also, this process could be used to generate biodiesel. This is admittedly a futuristic scenario, but it is the subject of intensive research and could make ethanol competitive with gasoline and make electricity competitive with other zero-carbon alternatives, especially when there is a price for avoiding CO_2 emissions.

The Global Warming Century

I take issue with the many scenarios that place us in the safe harbor of a hydrogen economy in just a decade or two. I do think such a transition may well be essential, for all the reasons I have described, but I think it will happen very differently from the way most analysts have suggested.

Here is my own scenario for the transition. Scenarios are not predictions; rather, they are credible and relevant narratives designed to challenge our thinking. This scenario assumes major technological advances but recognizes that the energy system changes slowly. As Shell noted in its most recent scenarios, "typically it has taken 25 years after commercial introduction for a primary energy form to obtain a 1% share of the global market."[69] As in Shell's, the environmental driver in this scenario is global warming. I draw on the most recent science and apply the lesson of the 1990s, when the hottest decade in 2,000 years drove most developed countries to a commitment to reduce their greenhouse gas emissions.

Although our scientific understanding of the climate will continue to improve in the coming years, this scenario plays out what a growing number of scientists see as the current warming trajectory. This scenario is called "The Global Warming Century." *It is a world driven by reaction to bad environmental news and good technolog-ical news.* Here are some of the highlights through 2050:

2000–2009: The hottest decade in 2,000 years. Many developed countries (other than the United States) begin making modest greenhouse gas reductions, creating a robust market for CO_2. Lack of leadership in the United States and the developing world stalls broader action on emissions. Growing R&D efforts and the prospect of major sales in Europe and Japan for climate-mitigating energy technologies bring about steady advances in fuel cells, renewables, energy efficiency, and hydrogen. CO_2 emissions in the United States rise by more than 10 percent, and in the world they rise by more than 15 percent.

2010–2019: The hottest decade in 2,000 years. The Intergovern-mental Panel on Climate Change (IPCC) raises its warming esti-mate in its fifth assessment, projecting that 2100 will be 3.5°F–12.5°F hotter than 1990. The IPCC acknowledges in 2016 that stabilization of CO_2 concentrations below 550 ppmv (twice preindustrial levels) by 2100 is "for all practical purposes, impos-sible." The National Academy of Sciences declares that 60 per-cent of the world's coral reefs will be lost by 2030 and that most of the rest are "unsavable, given current knowledge." The success Europe and Japan have in achieving small reductions at moder-ate costs inspires the first global climate treaty. Major developing nations refuse to accept absolute emissions reductions and insist that a nation's climate targets reflect cumulative emissions since 1900. The U.S. government refuses to accept a CO_2 reduction target before 2030. Solid oxide fuel cells and hybrid cars emerge as major commercial successes, delaying entry of PEM fuel cells into both the stationary and transportation markets. CO_2 emis-sions in the United States rise by another 10 percent, and world-wide they again rise by more than 15 percent. Global concentra-tions exceed 400 ppmv by the end of the decade.

2020–2029: The hottest decade in 2,000 years. The IPCC's seventh assessment projects that, by 2100, sea levels will be twelve to sixty inches higher than in 1990. After a piece of Antarctica the size of Connecticut breaks off in 2024, the National Academy of Sciences warns that disintegration of the entire Western Antarctic Ice Sheet now "appears likely within 200 years" and that a complete melting of the Greenland Ice Sheet is "virtually inevitable," which would raise sea levels more than twenty feet. In the United States, a major push for the most cost-effective climate solutions—energy efficiency, cogeneration, hybrid vehicles—leads to a peaking of U.S. primary energy demand in 2027. Fuel cell vehicles appear in a few fleets mandated by states such as California. Significant advances occur in hydrogen storage, coal and biomass gasification, and solar power. An electricity standard is proposed requiring that 50 percent of U.S. electricity be from renewables by 2040. The standard passes in 2025, after a "grand bargain" amendment allows CO_2 capture and storage to constitute as much as one-third of the requirement. A major effort is launched to characterize all significant potential geologic reservoirs for CO_2. Emissions of CO_2 in the United States remain flat, but global emissions rise by another 10 percent.

2030–2039: The hottest decade in 2,000 years. The IPCC's ninth assessment projects that 2100 will be 4.5°F–15°F warmer than 1990. The United States experiences its first 1,000-tornado month. In mid-decade, Hurricane Elisa hits Miami with winds raging at 200 miles per hour, killing 800 people and causing $100 billion in damage. In 2037, the National Academy of Sciences' Panel on Abrupt Climate Change, noting that the three previous years were a full 1°F warmer than the past decade, urges CO_2 stabilization at 650 ppmv within fifty years. Coal gasification together with carbon capture and storage (CCS) becomes cost-effective, with the resulting hydrogen-rich gas used to generate electricity in SOFC and gas turbine plants with efficiencies exceeding 80 percent. Global oil production peaks and starts to decline significantly in mid-decade as CO_2 restrictions limit development of unconventional resources. In 2038, China introduces the first 100 mpg fuel efficiency standard. Other countries quickly follow suit. CO_2 emissions in the United States drop by 20 percent, and global emissions are flat. Biofuels make up 25 percent of all vehicle fuels sold globally by decade's end.

2040–2049: The hottest decade in 2,000 years, another full 1°F warmer than the 2030s. The IPCC's eleventh assessment warns that "after five decades of anomalous behavior by the North Atlantic Oscillation, the thermohaline ocean current may be showing signs of shutting down." The United States launches a twenty-year effort to bio-refine all available biomass for biofuels production and for gasification, CCS, and hydrogen production. The last coal plant without CCS shuts down in 2048. By the end of the decade, most of the cars and light trucks on the road are hybrids (or hybrid diesel-electrics) running on an ethanol or biodiesel blend, and more than 10 percent of all new cars and light trucks are fuel cell vehicles. The hydrogen economy has begun.

CHAPTER 9

Hydrogen Partnerships and Pilots

The transition to a hydrogen economy will dramatically shift relationships in every sector of the economy. New and novel partnerships will be required everywhere—among governments, fuel providers, automobile manufacturers, consumer groups, and environmental groups. The transition calls for complicated decisions—which of the myriad options will we choose to produce, transport, store, and use hydrogen? During the many decades it will take us to solve the key technical, infrastructure, safety, and environmental issues, we will need to learn as much as possible about every option, to find out which are practical and which are pipe dreams. We must be ready to embrace new opportunities that we cannot even imagine today.

Federal and state governments have had limited and relatively unsuccessful experiences in trying to accelerate market adoption of alternative fuel vehicles. The private sector has fared no better because the path is difficult and many barriers reveal themselves only after technologies begin to be deployed. Because the transition to a hydrogen economy is so potentially important, a number of partnerships have already been created to launch hydrogen pilot projects around the world. This chapter focuses on two: those in Iceland and California.

The Iceland Experiment

The country poised to become the world's first hydrogen economy is an island nation in the North Atlantic Ocean the size of Kentucky, touching the Arctic Circle but warmed by the Gulf Stream. Iceland has more active volcanoes, hot springs, and geysers than any other area of its size in the world. It is a natural laboratory for learning how to tap into pollution-free heat and electricity. Its 280,000 people, less than 10 percent of Kentucky's population, are highly literate. Iceland has an infant mortality rate less than half that of the United States and longer life spans for both men and women. This is a unique combination of extraordinary natural resources, an educated populace, economic prosperity, and environmental awareness.

In particular, Iceland may be better suited to a near-term hydrogen transition than any other country because it has excess renewable energy. As discussed in the previous chapter, excess renewable energy is critical if hydrogen-powered transportation is to produce a net reduction in greenhouse gas emissions.

Iceland already uses renewable energy for electricity generation and space heating, so these sectors are nearly carbon-free, the only such example on the planet. The country's greenhouse gas emissions are produced by the transportation, fishing, and industrial sectors, each of which generates about 1 million metric tons of carbon dioxide (CO_2) equivalent per year.[1] By developing only a small fraction of its untapped renewable energy potential, Iceland could produce hydrogen that is nearly carbon-neutral and that could ultimately replace all the oil used for the country's transportation sector and huge fishing fleet. Iceland's CO_2 emissions would not be wholly eliminated, since emissions from industrial processes such as aluminum and ferrosilicon production would remain. Nevertheless, this plan would cut the country's fossil fuel use dramatically.

Iceland's Energy Resources

Greenland is where Max Pemberton based the hydrogen-powered pirate ship in his 1893 novel *The Iron Pirate*. The hydrogen is derived from coal, and one of the characters explains that Greenland "is one of the few comparatively uninhabited countries in the world where coal is to be had."[2] Ironically, today Greenland's neighbor is poised to be the first hydrogen-based economy precisely because it lacks major fossil fuel resources.[3]

Iceland has serendipitous geology—plentiful resources of inexpensive, clean hydropower and accessible geothermal energy. The country is located on a hot spot where the Eurasian and American continental plates meet and where hot magma from Earth's crust bubbles to the surface and forms active geothermal areas.[4] In fact, the Vikings who settled the island more than a thousand years ago named Reykjavík, the nation's capital, after *reykja,* the hot steam fumes billowing along the Icelandic horizon. Geothermal energy is derived from hot water or steam in the ground, often at great depths and pressures. Such high-temperature resources can power turbines, generating electricity. Moderate-temperature resources can directly heat buildings or provide process heat for industries.

Currently, geothermal energy is widely used in Iceland—a remarkable 90 percent of the country's buildings are heated by geothermal hot water or steam.[5] Overall, some 9 million megawatt-hours (mwh) of thermal energy is harnessed each year for heating and industrial uses.[6] Geothermal heating for buildings has the advantage of low cost; there is no need to buy or maintain heat pumps or furnaces or consume fuel of any kind. Iceland also has about 170 megawatts (mw) of geothermal electric generation capacity, providing more than 1.3 million mwh per year.[7] Iceland's geothermal resource base is considerable—some 30,000 mw, or 20-times the existing use.[8]

There are potential pitfalls. Geothermal energy may be extracted faster than it is recharged, so that many fields have limited lifetimes.[9] Energy production declines as the reservoir loses water, heat, or both. At the Geysers in northern California, for example, geothermal power production dropped rapidly in the 1980s, and slower declines have occurred in other locations. However, the vast differences in geology suggest that Iceland's geothermal reservoirs may not follow the same patterns as California's.

Research suggests that the current rate of geothermal energy use in Iceland is sustainable and could even be substantially increased.[10] The amount of energy stored in Iceland's unique bedrock is immense: 27 *trillion* MWh in the bedrock within three kilometers of the surface, of which 1 trillion is considered "usable."[11] Even if it were not continually recharged from below, this heat could be extracted from reservoirs and converted to energy for centuries to come at current usage levels. As it is, reservoir heat is replenished by volcanism (the movement of magma to the surface), advection (the movement of extremely hot water to the surface), and conduction (the transfer of heat through solid rock without any movement of material).

Iceland's potential to generate power from hydroelectric dams is also considerable because of the country's large area of mountainous terrain, glacial melt, and reliable precipitation. Hydroelectric plants in the country currently have a capacity of approximately 1,000 MW[12] and supply more than 6.8 million MWh per year of electricity.[13] In March 2003, Iceland approved plans for an Alcoa Inc. aluminum smelter and a 630 MW hydroelectric facility to provide power for this very electricity-intensive process.[14] This decision has encountered opposition from environmental groups because the resulting reservoir will flood land in an undeveloped wilderness area.

Because of such environmental concerns, Iceland may find it difficult to further develop its hydropower potential. Still, the current under-utilized capacity at hydroelectric plants would allow for significant hydrogen production, as explained by a study conducted by the World Wildlife Fund and the Iceland Nature Conservation Association:

> If the power plants providing electricity for the general market were to achieve the same load factor as the ones serving the power-intensive industries, then hydrogen could be produced during non-peak hours and stored until it is needed. Thus Iceland already has the electricity to replace 22% of the fossil fuels consumed by vehicles and vessels.[15]

Iceland also has the potential to develop wind power by installing coastal or offshore wind power facilities. The study noted that some 240 wind power plants could by themselves produce the electricity needed "to replace the fossil fuel from the transport and fisheries sector."[16]

Thus, Iceland has astonishing potential to expand its renewable energy production and produce carbon-neutral hydrogen. Some studies suggest that a mere 17 percent of the country's renewable energy potential has been developed.[17] The potential for renewable electricity has been estimated at 25 million to 50 million MWh per year for hydropower and geothermal energy.[18] Even the low end of the estimate would represent a tripling of current renewable energy capacity.

Iceland and the Energy Transition

In 1978, chemistry professor Bragi Árnason first proposed that Iceland develop a hydrogen economy. The idea was widely discussed and debated. Support grew rapidly in the 1990s, not only because of advances in fuel cell technology but also because of increased alarm about climate change and dependence on oil. In 1999, Shell, DaimlerChrysler, Norsk Hydro, the Icelandic holding company Vistorka hf (EcoEnergy), and other stakeholders formed the Icelandic Hydrogen and Fuel Cell Company, now Icelandic New Energy Ltd.[19] This consortium, with the backing of both the Icelandic government and the European Union, has begun to pioneer the hydrogen transition in Iceland.[20]

It is not just Iceland's renewable energy base that makes it an ideal laboratory for this experiment. Iceland is practically a city-state, with about 62 percent of the population living in or around the capital city, Reykjavík.[21] A hydrogen infrastructure on this remarkable little island can be established at relatively low cost—a few fueling stations in Reykjavík and on the roads around the island could easily serve most of the needs of the population. So the chicken-and-egg problem is far less daunting here. For example, on

April 24, 2003, Iceland opened the first public hydrogen filling station in the world, even though there are currently no privately owned hydrogen vehicles in the country.

The highly literate population of Iceland is widely supportive of the hydrogen initiative. A 2001 survey by the Sociological Institute of the University of Iceland found that 93 percent of Icelanders are positive about the idea of replacing traditional fossil fuels with hydrogen.[22] Their island is also prosperous and able to afford this investment; their standard of living is one of the highest in the world.

Iceland undertook two dramatic energy transitions in the twentieth century—shifting electricity generation from fossil fuel to hydroelectric power in the beginning of the century, and shifting space heating from fossil fuels to geothermal energy after World War II. Transportation is the last sector stuck on polluting fossil fuels.

Iceland is ripe for a third transition: To fuel its cars, buses, and fishing fleet, the country imports approximately 6 million barrels per year of petroleum.[23] At a cost of $30 per barrel, this amounts to $180 million per year. Gasoline and diesel fuel cost roughly $1 per liter ($3.80 per gallon).[24] Without indigenous sources of oil or other transportation fuels (aside from some landfill methane), the development of hydrogen becomes an attractive option.

Despite its carbon-free electricity sector and widespread use of geothermal heating, Iceland has high CO_2 emissions per capita. Typical developed countries emitted about 12 metric tons per capita in 1990, whereas Iceland emitted about 8.5 metric tons per capita.[25] All forms of energy, including renewables, affect their surroundings. Geothermal power may produce some emissions of CO_2, approximately 100 grams (g) per kilowatt-hour (kWh), roughly 30 percent of the emissions of a highly efficient combined cycle natural gas plant.[26] Increasing hydropower production may require additional dams, which carry an adverse environmental impact. On the other hand, Iceland has lots of unused capacity at existing hydroelectric power plants, which could be used to produce hydrogen in off-peak times. Revolutionizing the use of fuels for transportation is one of the last options left for Iceland to reduce its emissions levels significantly.

The Hydrogen Transition Plan

Icelandic New Energy has a six-phase plan for the transition to a hydrogen economy:[27]

- Phase 1 (under way): As noted earlier, a hydrogen fueling station was opened in April 2003, and three fuel cell buses, 4 percent of the city's bus fleet, are being introduced into Reykjavík. This phase, known as the ECTOS program (Ecological City Transport System), includes extensive analysis of the costs, benefits, time frame, and challenges to implementing a hydrogen transition in the transportation sector.[28]
- Phase 2: Gradually replace the Reykjavík city bus fleet (and perhaps other bus fleets) with proton exchange membrane (PEM) fuel cell buses.
- Phase 3: Introduce PEM fuel cell cars for private transportation.
- Phase 4: Demonstrate PEM fuel cell boats.
- Phase 5: Gradually replace the present fishing fleet with PEM vessels.
- Phase 6: Export hydrogen to the European continent.

The timing of later stages will depend on the progress of the early stages; the transition could be complete by 2030–2040.[29]

In the intermediate phases, which may well occur before a pure hydrogen fuel cell vehicle is practical, Iceland may begin the transition with *methanol-powered* PEM vehicles and vessels. The University of Iceland is examining the possibility of producing methanol (CH_3OH) from hydrogen combined with carbon monoxide (CO) or CO_2 from the exhaust of aluminum and ferrosilicon smelters. If most of the hundreds of thousands of tons of CO and CO_2 released by those smelters can be captured and then combined with hydrogen generated from electrolysis using renewable power, Iceland could cut its greenhouse gas emissions in half.[30]

The Global Iceland Hydrogen Partnership

In 2003, the nonprofit organization with which I work, the Global Environment and Technology Foundation (GETF), together with international experts in energy and sustainable development, approached Iceland's government to establish a Global Iceland Hydrogen Partnership. We received support from the Ministry for the Environment and the Ministries of Industry and Commerce. In this partnership, my colleagues and I will work with Icelandic New Energy, the government of Iceland, plus hydrogen researchers, stakeholders, and experts from around the world to assist Iceland in its hydrogen transition and to bring the lessons of the Iceland experience to other countries.

We are well aware of the many barriers to a hydrogen economy discussed elsewhere in this book, and it is too soon to know whether Iceland will be wholly successful in this transition. But given Iceland's size and commitment, we believe the infrastructure issues are manageable if the technical issues can be solved. Iceland is willing to invest in the production and storage technologies necessary to use hydrogen as a transportation fuel. Most important, the source of the hydrogen will be Iceland's own excess renewable energy, which means that their renewable power sources cannot be put to better uses, as they can in the United States and mainland Europe.

All of these factors serve to increase the likelihood that the hydrogen transition in Iceland can serve as a test bed and model for other nations.

The California Fuel Cell Partnership

California has the leading pilot program for hydrogen fuel cells in the United States. At the center of this effort is the California Fuel Cell Partnership (CaFCP), based in western Sacramento.[31] Its thirty members include most major automakers, a number of energy companies (BP, ChevronTexaco Corporation, ExxonMobil Corporation, and Shell Hydrogen), and several key government agencies. It was formed in 1999 to foster coordinated development and real-world testing of fuel cell vehicles.

California was chosen largely because that state has the strictest air quality standards in the nation and has been aggressively pushing zero emission vehicles (ZEVS). Since pure electric battery cars do not have proven marketplace success, fuel cell vehicles are seen as the logical next step in ZEV development. California has another advantage: Much of its electricity comes from renewable power sources and other low-carbon power generation, giving it about half the CO_2 emissions per kilowatt-hour of the nation as a whole. As discussed in Chapter 8, fuel cell vehicles will make the most sense as a greenhouse gas reduction strategy once the electric grid is virtually greenhouse-gas-free.

The CaFCP's stated goals are as follows:

- Demonstrate vehicle technology by operating and testing the vehicles under real-world conditions in California;
- Demonstrate the viability of alternative fuel infrastructure technology, including hydrogen and methanol stations;
- Explore the path to commercialization, from identifying potential problems to developing solutions; and
- Increase public awareness and enhance opinion about fuel cell electric vehicles, preparing the market for commercialization.[32]

By mid-2003, partnership members had helped put some three dozen fuel cell vehicles on the road in California—buses and light-duty vehicles—and dozens more were scheduled to be added in subsequent months. To begin servicing them, eight hydrogen fueling stations dotted the state from Los Angeles to Sacramento. A major goal of the CaFCP and other groups is to install more hydrogen fueling stations. In July 2003, Woody Clark, a senior policy advisor to the California Governor's Office, announced a plan to establish a "hydrogen freeway" that would provide fueling stations for hydrogen-powered vehicles at convenient locations up and down the state.

In March 2003 testimony before the House Science Committee, Alan Lloyd, 2003 chair of the partnership, explained, "The CaFCP maintains a 'fuel neutral' position regarding the choice of feed stock fuel for fuel cell vehicles. It's the common sense thing to do at this stage of exploration, in order to gain insight and experience with all potential fuels."[33] For instance, the hydrogen fueling station at the western Sacramento headquarters stores liquid hydrogen and then uses a compressor to raise the gas to high pressure for fueling. The station can also deliver methanol fuel for vehicles. Member companies also have technologies to generate hydrogen from fossil fuels and from electrolysis of water.

The partnership has funded detailed analysis of the barriers to and opportunities for moving from the pilot phase to full-scale commercialization, including an important 2001 report, "Bringing Fuel Cell Vehicles to Market: Scenarios and Challenges with Fuel Alternatives." The report notes that "a substantial risk premium may thus be applied by potential hydrogen infrastructure investors," in part because of the risk of stranded investment—the risk that early hydrogen infrastructure might be rendered unusable by future policy decisions or advances in technology (as discussed in Chapter 5). In this analysis, it takes ten years for investment in infrastructure to achieve a positive cash flow, and, to achieve this result, "significant technological advances will be required in reformers and electrolyzers, compressors, and overall systems integration as well as mass production methods for that equipment." Moreover, even a small tax on hydrogen appears to delay positive cash flow indefinitely.[34]

This report highlights some of the problems that will be faced in the future. A 2002 analysis of the difficulties faced in commercializing natural gas vehicles (NGVs) concluded: "The largest problem the NGV industry faced in Canada was a stalling in investment in public refueling facilities, which in turn retarded [vehicle] conversion sales. Investment in new refueling facilities stalled because existing stations did not build up sufficient load to make them profitable."[35] That's how the chicken-and-egg problem played out in Canada.

Conclusion

For hydrogen pilot programs, cost and profitability are secondary. The purpose is to gain experience with the new vehicles and their fuels. Pilots should focus on identifying areas where new research and new technologies are needed, creating a positive feedback loop that accelerates the development process. Pilots can also help create consensus about safety issues and about how new codes and standards should be written. If carefully designed and implemented over the next decade, a few pilots in states with various demographic, resource, and climate conditions could usefully expand the U.S. experience base.

States with pilot projects such as California should not be in a hurry to foster widespread commercialization, for many of the same reasons that the United States as a whole should move slowly. First, we still have many, many years of research and development before we will have solved key technology issues in hydrogen storage and transportation fuel cells. Second, we do not yet have any low-cost greenhouse-gas-free means of producing hydrogen; we have barely tapped the myriad strategies for reducing greenhouse gas emissions and oil imports that are likely to be far more cost-effective through at least 2030. Third, we must minimize the likelihood of stranded investments by not betting prematurely on technology that either is immature or cannot be easily modified for use with carbon-free hydrogen. Fourth, we should not accelerate the technology faster than our ability to ensure the public safety with well-designed codes and standards.

State-based programs such as California's have a number of specific challenges. They must demonstrate a long-term commitment to supporting vehicle infrastructure, or else their programs will meet the fate of NGVs and other alternative fuel vehicles. States that are today shifting money from clean energy funds into budget deficit reduction are sending a strong signal to the private sector that they cannot make the kind of long-term commitments that the shift to a hydrogen economy will require.

A key reason why an individual state cannot move too far ahead of the federal government in the hydrogen transition is that a state cannot solve the chicken-and-egg problem by itself, since the state cannot build the nationwide fueling infrastructure that will be required to induce automakers to build a significant number of these cars and encourage consumers to buy them. With that said, California, by virtue of its size and its historical leadership in the market for both cleaner cars and cleaner fuels, can play an important role in jump-starting and validating the hydrogen economy in the United States, much as Iceland can play the same role for the world. The hydrogen transition will take many, many decades, and we are fortunate that states such as California and countries such as Iceland are willing to go first and lead the way.

CONCLUSION

Choosing Our Future

This book has two primary themes. First, enabling the shift to a hydrogen economy may be one of the central tasks of the United States as we cope with the twenty-first century's major energy and environmental problems, especially global warming. Second, while widespread use of fuel cells for stationary power production seems likely post-2010, the transition to a transportation system based on hydrogen will take decades longer. Advances in technology alone will not prevent serious global warming nor deliver us to the Eden of a pollution-free hydrogen economy. Policy makers, businesses, and individuals all face tough choices.

We are close to the point at which even the best-case scenario for climate change will be fairly grim. Given the growing body of solid scientific evidence on global warming, and the remarkable consensus it has created in the scientific community, inaction during this decade is not mere procrastination. It represents a conscious choice to accept the risk of devastating and potentially irreversible climate change. We are in the doubly dangerous situation of facing global climate change that may be much more extreme and occur much more quickly than we expected just a few years ago. As the National Academy of Sciences explained in 2002, "the new paradigm of an abruptly changing climatic system has been well established by research over the last decade, but this new thinking is little known and scarcely appreciated in the wider community of natural and social scientists and policy-makers."[1]

A key job of government is first to identify threats to our security and well-being—especially plausible worst-case scenarios with catastrophic outcomes—and then to take steps to nip them in the bud. I believe the primary reason why we should pursue fuel cells and a hydrogen economy is to help respond to global warming. Yet hydrogen is no panacea. In the next three decades, it offers little or no prospect of helping the United States reduce its greenhouse gas emissions. Hydrogen will contribute significant reductions by 2050 *only* if we dramatically change the energy path we are now on. And so my overarching recommendation is as follows.

Take a long-term, conservative perspective on hydrogen. Overhyping the potential of hydrogen fuel cell vehicles will not bring them to the market sooner. More likely, it will create a backlash that will slow their ultimate market success—an important lesson from our humbling experience in the 1990s trying to speed the use of alternative fuel vehicles. Rather, we need to let research and development (R&D) do its steady work and, we hope, make major breakthroughs in all the key technologies—hydrogen production, storage, and infrastructure, as well as fuel cells and carbon sequestration.

Hydrogen-related analysis should be conservative in nature, given both the projected investment that will be required—hundreds of billions of dollars—and the risks to the planet of waiting for a deus ex machina that may not be practical for decades. Analysts should state clearly what is technologically and commercially possible today and, when discussing the future, be equally clear that projections are speculative and will require both major advances in technology and major government intervention in the marketplace. Analysis should treat the likely competition fairly: If major advances in cost reduction and performance are projected for hydrogen technologies, similar advances must be projected for hybrid vehicles, renewable biofuels, and the like. If hydrogen is being presented as a solution to problems such as global warming and dependence on imported oil, then the projected costs must be compared with those of the likely competition.

We also need to see that the likely avenue for the major introduction of fuel cells into the economy is, for the foreseeable future, *stationary* power production, including distributed generation running on natural gas. If carbon capture and storage (CCS) on a massive scale proves practical, then hydrogen could become a major energy carrier, but, again, it will most likely be used first for stationary power generation. The U.S. Department of Energy (DOE) has taken the position that it will not decide whether commercialization of fuel cell vehicles makes sense until 2015. Canadian engineer Geoffrey Ballard, founder of the leading proton exchange membrane (PEM) fuel cell company, said in June 2003, "The family-owned, garaged vehicle is the last vehicle that's going to get a fuel cell. Fuel cells are still 30 times the cost of what they need to be for the automotive market."[2]

We need continued expansion of R&D into key technologies. We also need pilot projects in order to gain practical experience with hydrogen-powered vehicles. In this country, pilots should be focused less on trying to speed early deployment of large numbers of fuel cell vehicles and more on trying to answer the key questions about storage, infrastructure, safety, and the like.

Most important, we need a variety of new government policies. A decade and a half of experience with countless *voluntary* emissions reduction programs run by the U.S. Environmental Protection Agency (EPA), the DOE, state governments, and nonprofit groups has proven that companies and individuals can reduce their pollution cost-effectively—but these voluntary programs have failed to stop the steady rise of U.S. greenhouse gas emissions. Only government action can reverse the growth of carbon dioxide (CO_2) emissions in order to buy time until hydrogen and other new technologies can help us reduce emissions sharply. And only government action can accelerate the development and deployment of greenhouse-gas-free power sources, which are key prerequisites for a hydrogen economy.

The following are specific strategies that can minimize the threat of global warming and maximize the chance that hydrogen will contribute to that effort.

Sharply increase research and development into clean energy technologies. The longer we delay taking direct steps to reduce our greenhouse gas emissions, the more we are gambling the future of the planet on multiple technology breakthroughs in fuel cells, hydrogen production, hydrogen storage, renewable energy, carbon capture and storage, energy efficiency, and on and on. But if we're going to bet our future on technology, let's give ourselves the best chance of winning. A $10 trillion economy with a $600 billion per year energy bill, and more than 5 billion metric tons of CO_2 emissions every year, *must* spend far more than the billion or so dollars the federal government now devotes to clean energy R&D. The increases in federal hydrogen and fuel cell R&D since 2000 have been welcome, but they have come largely at the expense of R&D on energy efficiency and renewable energy technologies, an unwise policy. Technological luck comes at a high price; we must more than double our federal energy R&D budget as rapidly as possible.

Start reducing CO_2 emissions now using low-cost strategies. Delay does not reduce the cost of stabilizing CO_2 emissions.[3] Quite the reverse. Delay increases costs for several reasons.

First, carbon-emitting products and facilities have a very long lifetime. Cars last 13 to 15 years. Coal plants can last 50 years. Commercial buildings and homes can last even longer. Replacing them is far more expensive than either building them smarter in the first place or, in the case of coal plants, not building most of them until we can minimize their emissions. Second, CO_2 lingers in the atmosphere, trapping heat, for more than a century. These twin facts create an urgency to avoid constructing another massive and long-lived generation of energy infrastructure that casts a carbon shadow into the 22ND century.

Third, as Prime Minister Tony Blair and Britain's Royal Commission on Environmental Pollution have said, we need to cut our emissions by more than 50 percent by 2050 in order to avoid a dangerous doubling of atmospheric CO_2 concentration. If we allow U.S. emissions to rise unchecked for the next two decades, that target will almost certainly be unachievable, and whatever target we do set will require faster and deeper—and hence costlier—changes in our energy infrastructure.

For anyone interested in speeding up the transition to a hydrogen economy, it is critical we quickly put in place a number of policies that do not appear, at first, directly related to hydrogen. As discussed in Chapter 8, using natural gas to make hydrogen for use in transportation makes no sense at all from the perspective of reducing net U.S. greenhouse gas emissions. Using carbon-free sources of electricity—such as renewable power or nuclear power or carbon capture and storage at a coal plant—to make hydrogen for transportation will not make sense until the electric grid has shifted to virtually carbon-free power. On our current energy path, that is extremely unlikely before 2050. In other words, *to the extent that global warming becomes the dominant energy policy driver in the near future, it won't make much sense to invest major public or private resources to install a nationwide hydrogen fueling infrastructure and to commercialize hydrogen vehicles before mid-century—unless we quickly change our current high-CO_2 energy path.*

Two low-cost policies can jump-start this shift: a renewable portfolio standard (RPS) and a cap on CO_2 emissions in the electricity sector. An RPS that requires 20 percent of U.S. electricity to be renewable by 2020 has very little net cost to the country and the huge benefit of reducing future natural gas prices.[4] Under such an RPS, electricity prices would be *lower* in 2020 than they are today, according to a 2001 study by the Energy Information Administration.[5] Caps on electric utility emissions of sulfur dioxide, oxides of nitrogen, and mercury have been proposed by many policy makers because they will dramatically improve air quality and save the lives of tens of thousands of Americans. Analysis by the EPA has shown that a relatively modest additional cap on grid CO_2 emissions—returning to 2001 levels by 2013—would add a mere two-tenths of a penny per kilowatt-hour in 2020, about 3 percent of electricity costs.[6]

Neither of these low-cost policies would have a dramatic effect on U.S. CO_2 emissions (and a tighter cap on carbon is certainly warranted). They would, however, do two vital things—get us off our unsustainable business-as-usual path and dramatically lower the cost of achieving the future deep reductions that are increasingly inevitable. They would also send a clear signal to the market that would spur significant investment in low-carbon technologies. Taken together these policies would accelerate the transition to a hydrogen economy by ten years or more. In particular, taking policy action this decade to set a carbon cap that kicks in during the next decade will be essential to spurring private sector investment in coal and biomass gasification as well as in carbon capture and storage. Such investment is essential if we are to launch a hydrogen economy before 2050, since the transition will not happen if the private sector cannot make money on it.

Begin a major national effort to encourage combined heat and power (CHP). To enable a hydrogen economy, we will need a fuel cell economy. How do we get there? The biggest potential market for stationary fuel cells is on-site cogeneration, or CHP. At the same time, CHP fueled by natural gas represents one of the lowest-cost strategies for reducing greenhouse gas emissions. Unfortunately, the barriers to rapid growth in CHP—and to all clean distributed generation technologies—remain high. The July 2000 report by the National Renewable Energy Laboratory discussed in Chapter 3 offered a "Ten-Point Action Plan for Reducing Barriers to Distributed Generation":

1. Adopt uniform technical standards for interconnecting distributed power to the grid.
2. Adopt testing and certification procedures for interconnection equipment.
3. Accelerate development of distributed power control technology and systems.
4. Adopt standard commercial practices for any required utility review of interconnection.
5. Establish standard business terms for interconnection agreements.

6. Develop tools for utilities to assess the value and impact of distributed power at any point on the grid.
7. Develop new regulatory principles compatible with distributed power choices in both competitive and utility markets.
8. Adopt regulatory tariffs and utility incentives to fit the new distributed power model.
9. Establish expedited dispute resolution processes for distributed generation project proposals.
10. Define the conditions necessary for a right to interconnect.[7]

Launch a major national effort to use electricity and natural gas more efficiently. We need to slow the growth of CO_2 emissions, sharply reduce the need for new coal-fired power, and free up inefficiently used natural gas for high-efficiency power generation (either combined cycle plants or CHP). *Energy efficiency remains the single most cost-effective strategy for minimizing CO_2 emissions.*[8]

Most buildings and factories can cut electricity consumption by 25 percent or more with rapid payback (less than four years). My 1999 book *Cool Companies* relates some one hundred case studies of companies that have done just that and made a great deal of money.[9] There are many reasons why most companies do not do what the best companies do, as explained in that book. The key point here is that we have more than two decades of experience with very successful state and federal energy efficiency programs. In short, we know what works.

Perhaps the most cost-effective federal strategy would be a matching program to co-fund state-based efficiency programs, with a special incentive to encourage states without an efficiency program to start one. This was a key recommendation of the Energy Future Coalition's End-Use Efficiency Working Group, a bipartisan effort to develop consensus policies in which I participated. Based on recent experience with state and utility efficiency programs, just $1 billion in annual federal matching funds for five years would, by 2015, cut projected U.S. electricity use by about 5 percent, or about 225 million megawatt-hours per year.[10] This policy would save consumers and businesses a whopping $15 billion in annual electricity bills. It would avoid the need for some 100 medium-sized (300 megawatt) power plants by 2015. *This simple strategy alone would avoid more than 100 million metric tons of CO_2 emissions per year.*

A natural gas efficiency strategy would be equally valuable. Again, most buildings and factories can cut natural gas consumption by 25 percent or more right now with rapid payback (less than four years). Removing barriers to CHP would be one element of this strategy. So would a major focus on more efficient use of steam. Steam is crucial for production in industries such as chemicals, food products, plastics, primary metals, pulp and paper, textiles, and petroleum refining. It is generated mainly by natural gas. Steam power represents a particularly large opportunity because it accounts for $20 billion per year of U.S. manufacturing energy costs and 40 percent of U.S. industrial CO_2 emissions.[11] Expanding state and federal efforts to use steam efficiently, such as the DOE's Best Practices Steam Program, would cut those numbers sharply.

These efficiency strategies, coupled with an effort to expand CHP, will also make the electric grid more resilient and thus less likely to suffer power blackouts such as the massive one that hit the United States and Canada in August 2003.

Phase in CO_2-related standards for cars and light trucks. We should aim for at least a 25 percent reduction in CO_2 emissions per mile for new vehicles by 2020. Transportation is the fastest-growing sector in terms of greenhouse gas emissions, and it is almost exclusively responsible for our large and growing dependence on imported oil. Absent such standards, emissions and imports will grow dramatically in the next two decades. A renewable fuels standard would also accelerate market entry of zero-carbon fuels such as cellulosic ethanol and hydrogen. General Motors said in 2003 that the promise of hydrogen cars justified delaying fuel-efficiency regulations.[12] But as we have seen, hydrogen offers little or no prospect of reducing U.S. greenhouse gas emissions for at least the next three decades and thus must not be used as an excuse for inaction.

These policies would have little net cost to the country but would carry huge benefits in reducing pollution and accelerating new technologies into the marketplace. Many of them are likely to be adopted by leading states such as California and New York. Unfortunately, these policies currently lack political support at the national level, and a number have been roundly rejected by the United States Congress, the president, or both. This rejection represents the choice to live with more than a doubling of heat-trapping CO_2 concentrations over preindustrial levels, which brings me to my final recommendation.

Prepare the public for the tough choices ahead. We are very likely entering a multi-decade period in which the recent heat waves and weather extremes around the globe will seem mild by comparison. By the middle of this century, the temperature may well start rising by 1°F per decade. Most of the world's coral reefs and their rich ecosystems probably cannot be saved. We face very real risks of catastrophic change in ocean circulation and sea level. Yet, from my perspective, rather than drawing attention to the growing dangers, the government, the media, and the environmental community have, if anything, been underplaying the risk.

Consider an article published in the *New York Times* in July 2003, "Records Fall as Phoenix All but Redefines the Heat Wave," highlighting daytime temperatures of 117°F and nighttime temperatures of 96°F—"the hottest night in Phoenix history." The article never mentions even the possibility that global warming might be part of the explanation or that scientists expect such heat waves to become both more commonplace and more severe.

Or consider an article in the *Washington Post* from the same month, "Coastal Louisiana Drowning in Gulf: Encroaching Salt Water Is Threatening the State's Economy and Homes." The article discusses a variety of reasons why Louisiana annually loses more than twenty-five square miles of coastland to the Gulf of Mexico, such as efforts to control the flow of the Mississippi River, but it never mentions even the possibility that climate change has contributed to the problem or that future sea level rise may undermine all efforts to find a long-term solution.

Such incomplete discussions leave the public ill prepared for what lies ahead and for what our nation will ultimately be called upon to do. Britain's Royal Commission on Environmental Pollution told Parliament in June 2000, "There is little public awareness or acceptance of the measures needed to accomplish sustained, deep reductions in greenhouse gas emissions."[13] And this from a country that had already publicly committed to reduce its greenhouse gas emissions and to adopt many policies currently unacceptable to U.S. government policy makers.

The longer we wait to act, and the more inefficient, carbon-emitting infrastructure that we lock into place, the more expensive and the more onerous will be the burden on all segments of society when we finally do act. While there are a variety of low-cost emissions reduction strategies available to us today, there is no cost-free technical fix to global warming poised just over the horizon. Most particularly, if hydrogen fuel cell cars are going to have a major impact on the nation's fossil fuel consumption and greenhouse gas emissions, the U.S. government will have to intervene in the market for cars and fuels far more intrusively than anything that has ever been attempted in the past.

Historians may write admiringly of the foresight of those who helped enable a hydrogen economy. But if we fail to act during *this decade* to reduce greenhouse gas emissions—especially if we do so because we have bought into the hype about hydrogen's near-term prospects—historians will condemn us because we did not act when we had the facts to guide us, and they will most likely be living in a world with a much hotter and harsher climate than ours, one that has undergone an irreversible change for the worse.

Acknowledgments

While the final judgments in the book are my own, I am exceedingly grateful to everyone who shared their ideas with me: John Atcheson, Andrew Browning, Stan Bull, Rodney Carlisle, Sean Casten, Tom Casten, Steve Chalk, Don Chen, Prashant Chintawar, Helena Chum, Mark Chupka, Gregg Cooke, William Cratty, Peter Dalpe, Kenneth Deffeyes, Raymond Drnevich, Alex Farrell, Stephen Folga, David Garman, Jerry Gillette, Sig Gronich, David Hamilton, David Hawkins, Steve Hester, Dale Heydlauff, Dennis Hughes, Tina Kaarsberg, George Kehler, David Keith, Tom Kreutz, Stephen Kukucha, John Lembo, James Lide, Michael Love, Andrew Lundquist, Lee Lynd, Alden Meyer, James McElroy, Peter Molinaro, William Morgan, Joan Ogden, Jim Olson, Michael Oppenheimer, Peter Pintar, David Redstone, Jennifer Rogers, Jim Rogers, Neil Rossmeissl, George Rudins, Barney Rush, Roger Saillant, Robert San Martin, Jennifer Schafer, Jeff Serfass, Phil Sharp, Dale Simbeck, Andrew Skok, David Scott Smith, Neil Stetson, George Thomas, Sandy Thomas, Mark Williams, John Wilson, William Wilson.

Special thanks go to those who reviewed all or part of the book: Brian Castelli, Jon Coifman, Bob Corell, Hank Habicht, Greg Kats, Jon Koomey, Mark Levine, Marianne Mintz, Fred Mitlitsky,

Pete O'Connor, Arthur Rosenfeld, Polly Shaw, K. R. Sridhar, Peter Teagan. I am also grateful to the National Hydrogen Association for the use of their library.

I have received a great deal of help from colleagues at the Center for Energy and Climate Solutions (www.cool-companies.org), a one-stop shop that provides companies and institutions with practical advice about the strategies and technologies available now for reducing emissions and cutting energy costs, including fuel cells. The Center and the World Wildlife Fund (WWF), in our Climate Savers Program, work with the world's most environmentally committed businesses to help them develop and adopt innovative climate and energy solutions. By 2010, the combined commitments of the first six Climate Savers companies will result in annual emissions reductions equivalent to more than 12 million metric tons of carbon dioxide.

The Center for Energy and Climate Solutions, which I founded in 1999, is a division of the Global Environment and Technology Foundation (GETF), a Virginia-based not-for-profit organization. Since 1988, GETF has brought industry, government, and communities together to address environmental challenges with innovative solutions. GETF is nonpartisan, with experts who have served in senior government positions in both Democratic and Republican administrations. In 2003, GETF launched the Global Hydrogen Initiative, which works with Iceland and other test-bed locations for hydrogen and fuel cells to document their activities and to assist them in their efforts. GETF staff members provided me with invaluable help in writing this book. I would especially like to thank Brian Castelli for sharing his decades of experience in all aspects of energy, Pete O'Connor for his solid research and analytical work, and Dallas Perkins for his help on research and references.

I also deeply appreciate and rely on the support and input from Hank Habicht and Greg Kats, my partners at the Capital E Group (www.cap-e.com), which we founded in 2001. Hank's passion for and expertise in the environment is matched only by his expertise in providing dispassionate counsel. Greg is one of the country's foremost experts on green buildings and on financing clean energy technologies. I have learned much from them.

Capital E is a premier provider of analytical and strategic consulting services to firms and investors in the clean energy industry. We work with a range of corporate and public clients to design and put in place energy systems that are cost-effective, reliable, and environmentally superior. Almost every building and factory that we look at can cut energy consumption (and hence greenhouse gas emissions) by 25 percent or more, with rapid payback. We also work with start-up companies, helping them raise funds, analyze markets, and find customers who will use their products. Among the companies we work with are ones that make fuel cells and other hydrogen-related technologies.

I would like to thank ScanSoft Inc. for its Dragon NaturallySpeaking voice dictation software. I have done most of my writing in the past several years using ScanSoft's software. Its best product yet, Dragon NaturallySpeaking 7, arrived just in time to enable me to double my productivity on this book.

I would like to thank the Energy Foundation for supporting the research and writing of this book and for supporting the Center from its inception.

Todd Baldwin, my editor at Island Press, made the entire process through the final product a very positive experience, and his editorial insight once again proved invaluable to the manuscript. He is the best book editor I've had the privilege of dealing with. I would also like to thank the entire staff of Island Press for their enthusiastic support of this book from the beginning. I am very appreciative of the work of Pat Harris, the most thorough of copyeditors.

I am indebted to my mother for applying her world-class language and editing skills to innumerable drafts.

Finally, special thanks go to my wife, Patricia Sinicropi, whose unflagging idealism and support provide me with unlimited inspiration every day, and without whose encouragement I would not have written this book.

Notes

Introduction

1. The SurePower case is based on Thomas Ditoro, "Banking on Fuel Cells to Supply Critical Loads," *Pure Power,* supplement to *Consulting-Specifying Engineer* (Fall 1998): 18–21; Dennis Hughes, William Cratty, and Arthur Mannion, personal communications; and information available at SurePower's Web site (www.hi-availability.com).

2. SurePower Press Release, "Major Credit Card Data Center Banking on Sure Power," May 24, 1999, www.hi-availability.com/nw_arch_press.html.

3. John Bockris, "The Origin of Ideas on a Hydrogen Economy and Its Solution to the Decay of the Environment," *International Journal of Hydrogen Energy* 27 (2002): 731–740.

4. David G. Hawkins, testimony before the U.S. House Committee on Energy and Commerce, Subcommittee on Energy and Air Quality, June 24, 2003, www.nrdc.org/globalWarming/tdh0603.asp. The issue of whether CO_2 could be captured from such plants is discussed in Chapters 7 and 8 of this book.

5. National Research Council (NRC), *Abrupt Climate Change: Inevitable Surprises* (Washington, DC: National Academies Press, 2002), www.nap.edu/books/0309074347/html.

6. Freedonia Group, "Hybrid & Competitive Automobile Powerplants," November 2002, distributed by Global Information Inc., Tokyo, www.the-infoshop.com/study/fd11850_hybrid.html.

7. See Tom Koppel, *Powering the Future: The Ballard Fuel Cell and the Race to Change the World* (Toronto: Wiley, 1999); and "Ballard Announces Five-Year Plan to Achieve Profitability, Lays Off 400, Raises New Cash," *Hydrogen & Fuel Cell Letter* 18 (January 2003).

8. See, for instance, Elizabeth Kolbert, "The Car of Tomorrow," *The New Yorker*, August 11, 2003, pp. 36–40 and James Mackintosh, "GM puts green gloss on image coloured by Hummer," *Financial Times*, October 7, 2003. In its comparative emissions

analysis, GM seems to be modeling a hybrid gasoline–electric vehicle that is only about 20% more efficient than a conventional internal combustion engine car, whereas the Toyota Prius is already 50% to 60% more efficient. See Keith Cole (director, Legislative and Regulatory Affairs, GM), "GM: Sustainable Mobility," Presentation to Fuel-cell Policy Roundtable, Dirksen Senate Office Building, Washington, DC, October 1, 2003.

Chapter 1: Why Hydrogen? Why Now?

1. George W. Bush, State of the Union address, January 28, 2003, www.whitehouse.gov/news/releases/2003/01/20030128-19.html.

2. Some experts insist on calling hydrogen an "energy carrier," somewhat like electricity, rather than a fuel, but I think either term is fine, "fuel" or "energy carrier."

3. Jim McElroy, personal communications.

4. Not unexpectedly, U.S. automakers wanted to focus on R&D with near-term payoffs and thus were not particularly supportive of increases in long-term research on PEM fuel cells. Many fossil fuel companies—including those in the natural gas industry—were equally unsupportive of PEM fuel cell research because they believed that hydrogen would inevitably come from non-fossil-fuel sources. As we will see, that is far from clear.

5. U.S. Department of Energy, Office of Energy Efficiency and Renewable Energy (EERE), "What Are People Saying About Hydrogen and Fuel Cells?" www.eere.energy.gov/hydrogenandfuelcells/feature.html.

6. Ibid.

7. Joseph J. Romm and Charles B. Curtis, "Mideast Oil Forever?" *Atlantic Monthly* 277, no. 4 (April 1996): 57–74, www.theatlantic.com/issues/96apr/oil/oil.htm.

8. David Garman (assistant secretary, Office of Energy Efficiency and Renewable Energy), speech given at Brownfields 2001 Conference, Chicago, IL, September 24, 2001, www.eere.energy.gov/brightfields/speech_09_01.html.

9. Romm and Curtis, "Mideast Oil Forever?"

10. Peter Schwartz and Doug Randall, "How Hydrogen Can Save America," *Wired*, no. 11.04 (April 2003), www.wired.com/wired/archive/11.04/hydrogen_pr.html.

11. U.S. Department of Energy, Energy Information Administration (EIA), *Annual Energy Outlook 2003* (Washington, DC: EIA, January 2003), table A19.

12. Barry C. Lynn, "Hydrogen's Dirty Secret," *Mother Jones* (May–June 2003), www.motherjones.com/news/outfront/2003/19/ma_375_01.html.

13. J. B. S. Haldane, "DAEDALUS, or Science and the Future" (paper read to the Heretics Society, Cambridge University, Cambridge, England, February 4, 1923), http://hem.passagen.se/aibpeter/energy/haldaneonenergy.html. See also Peter Hoffmann, *Tomorrow's Energy: Hydrogen, Fuel Cells, and the Prospects for a Cleaner Planet* (Cambridge, MA: MIT Press, 2001), pp. 27–51.

14. See, for instance, U.S. Department of Energy, Office of Energy Efficiency and Renewable Energy, "What is the hydrogen economy?" at www.eere.energy.gov/hydrogenandfuelcells (click on the words "hydrogen economy").

15. Dale Simbeck and Elaine Chang (SFA Pacific Inc.), "Hydrogen Supply: Cost Estimate for Hydrogen Pathways—Scoping Analysis" (Golden, CO: U.S. Department of Energy, National Renewable Energy Laboratory, July 2002).

16. See, for instance, JoAnn Milliken et al., "Grand Challenge for Basic and Applied Research in Hydrogen Storage" (U.S. Department of Defense presentation, Washington, DC, June 19, 2003), www.eere.energy.gov/hydrogenandfuelcells/pdfs/1_milliken_final.pdf.

17. Ulf Bossel and Baldur Eliasson (ABB Switzerland Ltd.), "Energy and the Hydrogen Economy," January 2003, p. 28, www.idatech.com/solutions/multi_fuel_solutions/Hydrogen%20Economy%20Report%202003.pdf.

18. Marianne Mintz et al. (Argonne National Laboratory), "Cost of Some Hydrogen Fuel Infrastructure Options" (presentation to the Transportation Research Board, January 16, 2002), www.transportation.anl.gov/pdfs/AF/224.pdf.

Chapter 2: Fuel Cell Basics

1. This history of fuel cells, as well as the discussion of technical characteristics of individual fuel cells, is based on Gregor Hoogers, ed., *Fuel Cell Technology Handbook* (Boca Raton, FL: CRC Press, 2002); *Fuel Cell Handbook*, 6th ed. (Morgantown, WV: EG&G Technical Services and Science Applications International Corporation for U.S. Department of Energy, National Energy Technology Laboratory, November, 2002), www.netl.doe.gov; the Smithsonian Institution's fuel cell history project, http://fuelcells.si.edu/index.htm; Peter Hoffmann, *Tomorrow's Energy: Hydrogen, Fuel Cells, and the Prospects for a Cleaner Planet* (Cambridge, MA: MIT Press, 2001); Sharon Thomas and Marcia Zalbowitz, *Fuel Cells—Green Power* (Los Alamos, NM: Los Alamos National Laboratory, 1999); U.S. Department of Energy, *Fuel Cell Report to Congress*, ESECS EE-1973, February 2003, www.eere.energy.gov/hydrogenandfuelcells/pdfs/fc_report_congress_feb2003.pdf; Energy Nexus Group, "Technology Characterization: Fuel Cells," report prepared for U.S. Environmental Protection Agency, April 2002, www.epa.gov/chp/pdf/EPA_Fuel_Cell_DraftCF.pdf; and K. R. Sridhar, Fred Mitlitsky, Roger Saillant, and Andrew Skok, personal communications.

2. Arno Evers, "Go to Where the Market Is! Challenges and Opportunities to Bring Fuel Cells to the International Market," *International Journal of Hydrogen Energy* 28 (2003): 725–733.

3. The figure is a slightly simplified version of one appearing at U.S. Department of Energy and U.S. Environmental Protection Agency, "How They Work: Fuel Cells," *Fuel Economy Guide*, www.fueleconomy.gov/feg/fcv_PEM.shtml.

4. Information is from UTC Fuel Cells' Web site, www.utcfuelcells.com/commercial/features.shtml.

5. The efficiency of a power-producing system like a fuel cell is essentially the useful work derived from the system divided by the heat content of the fuel put into the system. The heat content, or heating value, is the amount of energy that is released by burning the fuel. Confusion arises in comparing reported efficiencies because there are two categories of heating values, gross (higher) and net (lower). Burning a fuel produces some water vapor, which contains heat that could be released by condensation. If you include the energy in the water vapor, you get the higher (gross) heating value (HHV). If you don't, then the energy released is the lower (net) heating value (LHV).

The LHV can be 10% to 15% lower than the HHV for fuels such as natural gas, methanol, and hydrogen itself, for which the ratio is 15% (since burning hydrogen produces so much water). Manufacturers typically report efficiencies using LHV, since it is a smaller number and thus results in a higher efficiency, making their product look more attractive. The main point is to be consistent. Since LHV essentially represents the usable heat in a fuel, unless otherwise specified I will use LHV.

6. U.S. Department of Energy, Energy Information Administration (EIA), *A Look at Residential Energy Consumption in 1997* (Washington, DC: EIA, November 1999), p. 2, www.eia.doe.gov/pub/pdf/consumption/063297.pdf.

7. Peter Teagan (Arthur D. Little Inc.), "Fuel Cells for On-Site Generation" (presentation to ABST 2000, Washington, DC, June 2000).

8. *Fuel Cell Handbook,* 6th ed., pp. 1–12, 1–13.

9. David Milborrow and Lyn Harrison, "Hydrogen Myths and Renewables Reality," *Windpower Monthly* (May 2003): 47–53. Given that photovoltaics (PVs) provide electricity when it is most needed (during daytime peak electricity usage), it also seems unlikely that PVs would require expensive storage for the foreseeable future, except in certain remote applications. Batteries are currently used for such storage, and future reversible fuel cell systems would have to compete with their low cost, high reliability, and widespread market acceptance.

10. As reported in Gerald Cler, "The ONSI PC25 C Fuel Cell Power Plant," E Source Product Profile (Boulder, CO: E Source, March 1996).

11. See, for instance, U.S. Department of Energy, Office of Energy Efficiency and Renewable Energy, FY *2001 Progress Report for Fuel Cells for Transportation,*" December 2001, pp. 40–43, www.cartech.doe.gov/pdfs/FC/156.pdf.

Chapter 3: The Path to Fuel Cell Commercialization

1. See http://inventors.about.com/library/weekly/aacarsgasa.htm. The invention of the internal combustion engine is generally credited to Nikolaus August Otto in 1876, although there were numerous unsuccessful earlier efforts to design and/or build one.

2. International Energy Agency, "Retail Prices in Selected Countries in US Dollars/ Unit," www.iea.org/statist/keyworld2002/key2002/p_0505.htm. We have particularly low rates for industrial electricity.

3. Stacy C. Davis and Susan W. Diegel, *Transportation Energy Data Book,* 22nd ed. (Oak Ridge, TN: Oak Ridge National Laboratory, 2002), p. 5-2.

4. See, for instance, Michael E. Porter, *The Competitive Advantage of Nations* (New York: Free Press, 1990).

5. U.S. Department of Energy, *Fuel Cell Report to Congress,* ESECS EE-1973, February 2003, p. vi, www.eere.energy.gov/hydrogenandfuelcells/pdfs/fc_report_ congress_ feb2003.pdf.

6. Shell International, Global Business Environment, "Energy Needs, Choices, and Possibilities: Scenarios to 2050," 2001, www.shell.com/static/media-en/downloads/ 51852.pdf.

7. The SurePower case is based on Thomas Ditoro, "Banking on Fuel Cells to Supply Critical Loads," *Pure Power,* supplement to *Consulting-Specifying Engineer* (Fall 1998): 18–21; Dennis Hughes, William Cratty, and Arthur Mannion, personal communications; and information available at SurePower's Web site (www.hi-availability.com).

8. For a more detailed discussion of availability and other issues related to high-reliability generation, see Chris Robertson and Joseph Romm, "Data Centers, Power, and Pollution Prevention: Design for Business and Environmental Advantage" (Arlington, VA: Center for Energy and Climate Solutions, June 2002), www.coolcompanies.org/images/DataCenterJune.pdf.

9. Peter Huber and Mark Mills, "Dig More Coal—the PCs Are Coming," *Forbes* (May 31, 1999). See also Mark Mills, *The Internet Begins with Coal: A Preliminary Exploration of the Impact of the Internet on Electricity Consumption* (Arlington, VA: Greening Earth Society, June 1999).

10. For instance, Bank of America Securities concluded in a major 200-page report in June 2000, "Internet power will account for 40% of total electrical load growth." See Bank of America Securities, "The Power of Growth," June 2000, p. 12. Also, "Electricity is a conservative play on the Internet," reported Salomon Smith Barney analysts in 2000, "and all Internet technology is plugged into an electrical outlet." Cited in David Wessel, "Bold Estimate of Web's Thirst for Electricity Seems All Wet," *Wall Street Journal* (December 5, 2002), http://online.wsj.com/article/0,,SB10390403142288I3793.djm,00.html.

11. Jon Koomey et al., "Initial comments on *The Internet Begins with Coal*," memorandum to Skip Laitner of the U.S. Environmental Protection Agency's Office of Atmospheric Programs, Lawrence Berkeley National Laboratory, Berkeley, CA, December 1999. LBNL maintains a Web site devoted to debunking this myth, with links to all relevant articles and studies (http://enduse.lbl.gov/Projects/infotech.html).

12. The head of the U.S. Department of Energy's Energy Information Administration testified in February 2000, "From 1985 to 1995, retail electricity sales grew at a rate of 2.6% per year. . . . Since 1995, the use of the Internet has increased dramatically, yet retail electricity sales have grown by 2.1% per year." U.S. Department of Energy, Energy Information Administration, "Statement of Jay Hakes before the Subcommittee on National Economic Growth," U.S. House of Representatives, Committee on Government Reform, Washington, DC, February 2, 2000.

13. Joseph Romm, Arthur Rosenfeld, and Susan Herrmann, "The Internet Economy and Global Warming: A Scenario of the Impact of E-commerce on Energy and the Environment" (Arlington, VA: Center for Energy and Climate Solutions, December 1999), www.coolcompanies.org/energy/paper1.cfm. See also Joseph Romm, "The Internet and the New Energy Economy," *Resources, Conservation, and Recycling* 36, no. 3 (October 2002): 197–210.

14. Kurt W. Roth, Fred Goldstein, and Jonathan Kleinman, *Energy Consumption by Office and Telecommunications Equipment in Commercial Buildings* (Cambridge, MA: Arthur D. Little, January 2002), www.eere.energy.gov/buildings/documents/pdfs/office_telecom-vol1_final.pdf. The study noted that Mills' estimates of annual energy consumption exceeded that of Arthur D. Little Inc. (and "all other researchers") for "all the equipment types considered." Mills exceeded Arthur D. Little's estimate of the energy consumed by more than a factor of six.

15. Walter S. Baer, Scott Hassell, and Ben Vollaard, *Electricity Requirements for a Digital Society* (Santa Monica, CA: Rand, 2002).

16. Wessel, "Bold Estimate."

17. The Arthur D. Little report projected the likely energy consumption of informa-

tion technology (IT) equipment to the year 2010 under a number of scenarios. Even in the scenario with the largest growth in IT equipment, it consumes only about "3.5% of the projected national electricity consumption in Y2010." In other scenarios, the figure is less than 2 percent. Rand was unable to find a "set of plausible assumptions" in which the figure hit 10 percent by 2021, and it concluded, "In none of our 2021 scenarios does this percentage exceed 5.5 percent." Roth, Goldstein, and Kleinman, *Energy Consumption by Office and Telecommunications Equipment;* Baer, Hassell, and Vollaard, *Electricity Requirements for a Digital Society.*

18. For further discussion, see Robertson and Romm, "Data Centers."

19. SurePower maintains a fuel-cell-based product called Ultra-Green for smaller data centers "from 200 kilowatts to 2.0 megawatts" (www.hi-availability.com/pr_ultra_green.html).

20. Paul Lancaster, quoted in "Defining Commercialization by Design, Price Point, and Volume," *Fuel Cell Industry Report* 3, no. 10 (October 2002): 1.

21. Tina Kaarsberg et al. (U.S. Department of Energy), "Market Perspectives of Micro CHP" (paper presented December 2000, Washington, DC), and Kaarsberg, personal communications; Tom Kreutz and Joan Ogden, *Assessment of Hydrogen-Fueled Proton Exchange Membrane Fuel Cells for Distributed Generation and Cogeneration,* final report to the U.S. Department of Energy (Princeton, NJ: Princeton University, Center for Energy and Environmental Studies, October 2000); Tom Kreutz and Joan Ogden, "Prospective Performance and Economics of Residential Cogeneration Using Natural Gas–Fueled PEM FC Power Systems" (paper presented at the 11th Annual US Hydrogen Meeting, Vienna, VA, March 2, 2000), and Kreutz, personal communications; C. E. (Sandy) Thomas, Brian D. James, and Franklin D. Lomax Jr. (Directed Technologies Inc., Arlington, VA), "Analysis of Residential Fuel Cell Systems and PNGV Fuel Cell Vehicles," in *Proceedings of the 2000 DOE Hydrogen Program Review,* www.eere.energy.gov/hydrogenandfuelcells/pdfs/28890mm.pdf; and Drew Ronneberg, "A Fuel Cell in Every Home?" (Washington, DC: Capital E Group, 2002).

22. Kaarsberg et al., "Market Perspectives." In general, the best way to design a cogeneration system to optimize both economic return and emissions reductions is to design the system around the thermal needs of a building or factory, maximizing use of the waste heat. Unfortunately, this is not how most companies are thinking about residential PEMs. Such an optimally designed home fuel cell would provide only the so-called base-load electric power of the home, and the homeowner would still need to rely on the local utility for peak power.

23. Stephen Kukucha, personal communication.

24. Plug Power, "Plug Power Introduces First Fuel Cell Product for the Telecommunications Market," press release, June 3, 2003, www.plugpower.com; Roger Saillant, personal communication.

25. Kreutz and Ogden, "Prospective Performance and Economics of Residential Cogeneration"; Kreutz and Ogden, *Assessment of Hydrogen-Fueled Proton Exchange Membrane Fuel Cells.*

26. Tom Kreutz, personal communications.

27. George Erdmann, "Future Economics of the Fuel-Cell Housing Market," *International Journal of Hydrogen Energy* 28 (2003): 685–694.

28. The Starwood case study is based on Rebecca Oliva, "Alternative Energy Solutions," *Hotels* (July 2002), www.hotelsmag.com/0702/0702tech.html; John Lembo, "Starwood's Strategic Energy Management Initiative," *HPAC Engineering* (February 2003), www.hpac.com/member/feature/2003/0302/0302lembo.htm; and John Lembo, personal communications.

29. The Dow-GM case is based on "Dow Plans to Use GM Fuel Cells in World's Largest Fuel Cell Transaction," joint corporate press release, Washington, DC, May 7, 2003; and George Kehler and Peter Molinaro, personal communications.

30. For a large chemical manufacturer such as Dow, with access to huge quantities of natural gas at relatively low industrial prices, cogeneration costs can be as low as two to three cents per kilowatt-hour for electricity.

31. At the state level, however, there is much variation. In the Midwest, most of the electricity comes from coal, and CO_2 emissions per unit of electricity (per kilowatt-hour, for example) can be almost twice as much as the national average. In California, the electric system is much cleaner, and the CO_2 emissions per kilowatt-hour are closer to that of a combined cycle gas turbine plant, which can have an efficiency of more than 55 percent.

32. If the United States were to restrict CO_2 emissions in the electric utility sector, however, new distributed energy systems would displace the most carbon-intensive power plants, which are coal fired.

33. Ken Hydzik (Avista Labs), "Final Project Report for Geiger Field Installation," 242ND Combat Communication Squadron, Geiger Field, WA, August 18, 2003.

34. For details, see Joseph Romm, "Greenhouse Gas Analysis of the SurePower System at the First National Bank of Omaha," June 1, 1999, www.hi-availability.com/pdf/greenhouse.pdf. Omaha's electricity has slightly higher CO_2 intensity than that of the U.S. electric grid.

35. Another strategy for companies interested in pollution-free power is to switch to a "green power" provider that delivers certified renewable electricity. This will often be the low-cost option.

36. Peter Teagan (Arthur D. Little Inc.), "Fuel Cells for On-Site Generation" (presentation to ABST 2000, Washington, DC, June 2000).

37. ONSITE SYCOM Energy Corporation, "The Market and Technical Potential for Combined Heat and Power in the Commercial/Institutional Sector," report prepared for U.S. Department of Energy, Energy Information Administration (Washington, DC: ONSITE SYCOM Energy Corporation, January 2000), www.eere.energy.gov/der/chp/pdfs/eiacom.pdf.

38. Energy Nexus Group, "Technology Characterization: Gas Turbines," report prepared for U.S. Environmental Protection Agency, February 2002, www.epa.gov/chp/pdf/EPA_Gas_Turbines_final_April_29_02.pdf.

39. Energy Nexus Group, "Technology Characterization: Reciprocating Engines," report prepared for U.S. Environmental Protection Agency, February 2002, www.epa.gov/chp/pdf/EPA_RecipEngines_final_5_16_02.pdf.

40. Energy Nexus Group, "Technology Characterization: Steam Turbines," report prepared for U.S. Environmental Protection Agency, March 2002, www.epa.gov/chp/pdf/EPA_Steam_TurbinesCF.pdf.

41. Companies focused on reducing net CO_2 emissions will take a longer-term perspective to make sure they choose the approach that minimizes costs *and* emissions.

42. Sean Casten, personal communications.

43. R. Brent Alderfer, Thomas J. Starrs, and M. Monika Eldridge, "Making Connections: Case Studies of Interconnection Barriers and Their Impact on Distributed Power Projects" (Golden, CO: U.S. Department of Energy, National Renewable Energy Laboratory, July 2000), www.nrel.gov/docs/fy00osti/28053.pdf.

44. FuelCell Energy investors conference call, June 3, 2003.

Chapter 4: Hydrogen Production

1. See, for instance, Dale Simbeck (SFA Pacific Inc.), "Coal—Bridge to the Hydrogen Economy" (presentation to the Eighteenth Annual International Pittsburgh Coal Conference, Newcastle, New South Wales, Australia, December 4, 2001), www.engr.pitt.edu/pcc/2001plenary/SimbeckCoaltoH2revised.pdf; and William Wilson (Cambrensis Ltd., www.cambrensis.org), "The Hydrogen Economy—a View from Britain and Europe" (presentation to the National Hydrogen Association's Fourteenth Annual U.S. Hydrogen Conference and Hydrogen Expo, Washington, DC, March 4–6, 2003).

2. This brief history is based on John Rigden, *Hydrogen: The Essential Element* (Cambridge, MA: Harvard University Press, 2002), which is also the source of all the quotes in this chapter by scientists; Peter Hoffmann, *Tomorrow's Energy: Hydrogen, Fuel Cells, and the Prospects for a Cleaner Planet* (Cambridge, MA: MIT Press, 2001); and U.S. Department of Energy, Office of Energy Efficiency and Renewable Energy, Hydrogen, Fuel Cells, and Infrastructure Technologies Program, "Hydrogen: Frequently Asked Questions," www.eere.energy.gov/hydrogenandfuelcells/hydrogen/faqs.html.

3. Rigden's full quote is, "With one exception—the helium atom—hydrogen is the mother of all atoms and molecules." Rigden, *Hydrogen: The Essential Element*.

4. Jules Verne, *The Mysterious Island*, www.online-literature.com/verne/mysteriousisland/33/. Emphasis added.

5. Max Pemberton, *The Iron Pirate* (London: Greycoine Book Manufacturing Company, n.d.).

6. The hydrogen facts in this section are from U.S. Department of Energy, *Proceedings, National Hydrogen Energy Roadmap Workshop, Washington, DC, April 2–3, 2002*, www.eere.energy.gov/hydrogenandfuelcells/pdfs/workshop_proceedings.pdf; Gregor Hoogers, ed., *Fuel Cell Technology Handbook* (Boca Raton, FL: CRC Press, 2002); U.S. Department of Energy, "Hydrogen: Frequently Asked Questions"; Dale Simbeck (SFA Pacific Inc.), "CO_2 Capture and Storage—the Essential Bridge to the Hydrogen Economy" (paper presented at the Sixth International Conference on Greenhouse Gas Control Technologies [GHGT-6], Kyoto, Japan, October 1–4, 2002); and U.S. Department of Energy, Office of Fossil Energy, "Hydrogen and Other Clean Fuels," www.fe.doe.gov/programs/fuels/.

7. U.S. Department of Energy, Office of Energy Efficiency and Renewable Energy, Hydrogen, Fuel Cells, and Infrastructure Technologies Program, "Hydrogen: Frequently Asked Questions," www.eere.energy.gov/hydrogenandfuelcells/hydrogen/faqs.html#year.

8. U.S. Department of Energy, *Proceedings, National Hydrogen Energy Roadmap Work-

shop, p. 7., and *Fuel Cell Handbook*, 6th ed. (Morgantown, WV: EG&G Technical Services and Science Applications International Corporation for U.S. Department of Energy, National Energy Technology Laboratory, November, 2002), www.netl.doe.gov, pp. 10–32.

9. Stacy C. Davis and Susan W. Diegel, *Transportation Energy Data Book*, 22nd ed. (Oak Ridge, TN: Oak Ridge National Laboratory, 2002), pp. 2-9, 2-11.

10. Natural gas isn't pure methane but a naturally occurring mixture of gases, primarily (about 95 percent) methane, but also including a variety of other gases that varies by region. For a typical breakdown, see Union Gas Limited, *Chemical Composition of Natural Gas*, www.uniongas.com/NaturalGasInfo/AboutNaturalGas/composition.asp.

11. U.S. Department of Energy, "Hydrogen and Other Clean Fuels"; Joan Ogden, "Review of Small Stationary Reformers for Hydrogen Production," report to the International Energy Agency (Princeton, NJ: Princeton University, Center for Energy and Environmental Studies, March 2001), www.princeton.edu/~cmi/research/papers/review.pdf; Marc Jensen and Marc Ross, "The Ultimate Challenge: Developing an Infrastructure for Fuel-Cell Vehicles," *Environment* 42, no. 7 (September 2000): 10–22.

12. U.S. Department of Energy, "Hydrogen and Other Clean Fuels." See also Dale Simbeck and Elaine Chang (SFA Pacific Inc.), "Hydrogen Supply: Cost Estimate for Hydrogen Pathways—Scoping Analysis" (Golden, CO: U.S. Department of Energy, National Renewable Energy Laboratory, July 2002). Simbeck and Chang use a natural gas price of $3.50 per million Btu (higher heating value, or HHV) for centralized production and $6 per million Btu for forecourt production.

13. Simbeck and Chang, "Hydrogen Supply," pp. 23–24.

14. This is true even assuming that fuel cell vehicles are twice as efficient as internal combustion engine vehicles.

15. This does not include the full life-cycle CO_2 emissions from the burning of gasoline, which includes the impact of drilling, transporting, and refining the oil and is typically calculated to increase emissions by about 20 percent. Nor does it include the full life-cycle CO_2 emissions from the elements of the hydrogen infrastructure. The full "apples-to-apples" comparison is considered in Chapter 8.

16. Although hydroelectric power is far and away the dominant source of renewable electricity, there is little prospect of a significant increase in U.S. hydropower production, so it does not represent a major potential source of new zero-carbon hydrogen.

17. Partial oxidation of a hydrocarbon results in CO and hydrogen (and heat), so it is typically followed with a shift reaction and hydrogen purification. Such a system has an efficiency in converting gasoline into usable hydrogen of 70 to 80 percent or more. It potentially has a faster response time and is more compact than a steam reformer, so it is more suitable for reforming on board a vehicle. In autothermal reformers, steam reforming is combined with partial oxidation, and the latter reaction provides heat for the former, thereby increasing efficiency.

18. Information from *Hydrogen & Fuel Cell Investor's Newsletter* 7, no. 13 (May 25, 2003); information from www.nuvera.com; and Prashant Chintawar, personal communication.

19. California Fuel Cell Partnership (CAFCP), "Bringing Fuel Cell Vehicles to Market: Scenarios and Challenges with Fuel Alternatives," report prepared by Bevilacqua Knight Inc., October 2001, www.fuelcellpartnership.org/documents/ScenarioStudy_VI-1.pdf. A CAFCP position paper was released with this report (www.cafcp.org/documents/

PositionPaperFinal.PDF). Typically, gasoline reformers prefer fuel with very low levels of sulfur. Some fuel providers have already proposed producing very low-sulfur gasoline.

20. *Hydrogen & Fuel Cell Investor's Newsletter* 7, no. 13 (May 25, 2003).

21. The methanol section is based on U.S. Environmental Protection Agency, Office of Transportation and Air Quality, Transportation and Regional Programs Division, "Clean Alternative Fuels: Methanol," March 2002, www.epa.gov/otaq/consumer/fuels/altfuels/methanol.pdf; Malcolm Pirnie Inc., "Evaluation of the Fate and Transport of Methanol in the Environment," report prepared for the American Methanol Institute, Washington, DC, January 1999, www.methanex.com/fuelcells/archivedreports/evaluation.pdf; Breakthrough Technologies Institute, "Beyond the Internal Combustion Engine: The Promise of Methanol Fuel Cell Vehicles," report prepared for the American Methanol Institute, Washington, DC, February 2001, www.methanex.com/fuelcells/archivedreports/meohfcvs.pdf; and information available at www.methanol.org.

22. CaFCP, "Bringing Fuel Cell Vehicles to Market."

23. My company, the Capital E Group, works with a company that makes methanol from landfill gas.

24. Breakthrough Technologies Institute, "Beyond the Internal Combustion Engine."

25. The American Coal Foundation, "FAQ and Coal Quiz," www.acf-coal.org/pages/FAQ.html.

26. U.S. Department of Energy, Office of Fossil Energy, "Hydrogen from Coal Research," www.fe.doe.gov/programs/fuels/hydrogen/hydrogen-from-coal.shtml.

27. U.S. Department of Energy, Office of Fossil Energy, "FutureGen: A Sequestration and Hydrogen Research Initiative," fact sheet, February 2003, www.energy.gov/engine/doe/files/import/FutureGenFactSheet.pdf.

28. D. Gray and G. Tomlinson, "Hydrogen from Coal," Mitretek Technical Paper MTR 2002-31, report for U.S. Department of Energy, National Energy Technology Laboratory (NETL), July 2002, www.netl.doe.gov/coalpower/gasification/pubs/pdf/HYDROGEN%20FROM%20COAL4.pdf.

29. Dale Simbeck, "Gasification Repowering: The Innovative Option for Old Existing Coal-Fired Power Plants" (presentation to the eleventh annual National Energy Modeling System/Annual Energy Outlook Conference, Washington, DC, March 18, 2003). Simbeck proposes retrofitting existing coal plants with gasification and sequestration technology rather than building entirely new plants.

30. The biomass discussion is based on Pamela L. Spath et al., "Update of Hydrogen from Biomass: Determination of the Delivered Cost of Hydrogen" (Golden, CO: U.S. Department of Energy, National Renewable Energy Laboratory, April 2000, revised July 2001, www.decisioneering.com/articles/download/biomass_to_hydrogen.pdf); Y. D. Yeboah et al., "Hydrogen from Biomass for Urban Transportation," in *Proceedings of the 2002 U.S. DOE Hydrogen Program Review* (Golden, CO: U.S. Department of Energy, National Renewable Energy Laboratory), www.eere.energy.gov/hydrogenandfuelcells/pdfs/32405a10.pdf; Simbeck and Chang, "Hydrogen Supply"; and Helena Chum, personal communications.

31. Biomass, whether used for energy purposes or not, will typically decay, giving off greenhouse gases. If the biomass can be burned cleanly (or converted to hydrogen), it can serve to offset the combustion of fossil fuels.

32. Lester B. Lave, W. Michael Griffin, and Heather MacLean, "The Ethanol Answer to Carbon Emissions," *Issues in Science and Technology* (Winter 2001), http://search.nap.edu/issues/18.2/lave.html.

33. One possible way to cut costs is to pyrolyze biomass to produce bio-oil in facilities near the source of the biomass and then truck the oil to hydrogen-producing facilities closer to fueling stations.

34. Spath et al., "Update of Hydrogen from Biomass."

35. The nuclear energy discussion is based on U.S. Department of Energy, Office of Nuclear Energy, Science and Technology, "Nuclear Hydrogen Initiative," January 2003, http://nuclear.gov/infosheets/Hydrogen%20J.pdf; Ken Schultz, "Efficient Production of Hydrogen from Nuclear Energy" (presentation to the California Hydrogen Business Council, June 27, 2002), www.ch2bc.org/General%20Atomics/NuclearH2-27June02.pdf.

36. L. C. Brown et al., "High Efficiency Generation of Hydrogen Fuels Using Nuclear Energy," a Nuclear Energy Research Initiative (NERI) Project for the U.S. Department of Energy, summarized by Ken Schultz for the Hydrogen and Fuel Cells Annual Review, May 6, 2002, www.eere.energy.gov/hydrogenandfuelcells/pdfs/32405d.pdf. See also L. C. Brown et al., "Nuclear Production of Hydrogen Using Thermochemical Water-Splitting Cycles," http://web.gat.com/pubs-ext/MISCONF02/A23944abs.pdf.

37. John Deutch and Ernest Moniz et al., *The Future of Nuclear Power*, Massachusetts Institute of Technology, Cambridge, MA, July 2003, http://web.mit.edu/nuclear-power/.

38. M. G. J. Janssen et al., "Biohydrogen 2002," *International Journal of Hydrogen Energy* 27 (2002): 1123–1124. See also Debabrata Das and T. Nejat Veziroglu, "Hydrogen Production by Biological Processes: A Survey of Literature," *International Journal of Hydrogen Energy* 26 (2001): 13–28.

39. K. R. Sridhar, personal communication.

Chapter 5: Key Elements of a Hydrogen-Based Transportation System

1. Bernard Bulkin, presentation to the National Hydrogen Association's Fourteenth Annual U.S. Hydrogen Conference and Hydrogen Expo, Washington, DC, March 4–6, 2003.

2. Marianne Mintz et al. (Argonne National Laboratory), "Cost of Some Hydrogen Fuel Infrastructure Options" (presentation to the Transportation Research Board, January 16, 2002), www.transportation.anl.gov/pdfs/AF/224.pdf.

3. Marc Melaina, "Initiating Hydrogen Infrastructures," *International Journal of Hydrogen Energy* 28 (2003): 743–755.

4. The discussion of alternative fuel vehicles is based on U.S. General Accounting Office (GAO), *Energy Policy Act of 1992: Limited Progress in Acquiring Alternative Fuel Vehicles and Reaching Fuel Goals*, GAO/RCED-00-59 (Washington, DC: GAO, February 2000), www.ccities.doe.gov/pdfs/GAOprogress.pdf, and Brian Castelli, personal communication.

5. Peter Flynn, "Commercializing an Alternate Vehicle Fuel: Lessons Learned from Natural Gas for Vehicles," *Energy Policy* 30, no. 7 (2002): 613–619.

6. U.S. Department of Energy, *Fuel Cell Report to Congress*, ESECS EE-1973, February 2003, www.eere.energy.gov/hydrogenandfuelcells/pdfs/fc_report_congress_feb2003.pdf.

7. The NASA discussion is from National Aeronautics and Space Administration, Kennedy Space Center, "NSTS Shuttle Reference Manual," http://science.ksc.nasa.gov/shuttle/technology/sts-newsref/et.html; Kelly Young, "NASA's Dangerous Cargo: Volatile Fuels Traverse State," *Florida Today* (July 16, 2003), www.floridatoday.com/news/space/stories/2002a/071702fuel.htm; and Aaron Hoover, "Florida Universities Solving NASA's Hydrogen Problem," *University Science News*, April 19, 2002, http://unisci.com/stories/20022/0419026.htm.

8. Gerd Arnold (General Motors Corporation, Global Alternative Propulsion Center), "Advanced Hydrogen Storage Technologies," www.gmfuelcell.com/w_shoop/pdf/Gerd%20Arnold(E).pdf.

9. George Thomas and Jay Keller, "Hydrogen Storage—Overview" (presentation to the H_2 Delivery and Infrastructure Workshop, Washington, DC, May 7–8, 2003), www.eere.energy.gov/hydrogenandfuelcells/pdfs/bulk_hydrogen_stor_pres_sandia.pdf.

10. JoAnn Milliken (U.S. Department of Energy), "Hydrogen Storage Activities under the Freedom Car & Fuel Initiative" (presentation to the National Hydrogen Association's Fourteenth Annual U.S. Hydrogen Conference and Hydrogen Expo, Washington, DC, March 4–6, 2003).

11. Ulf Bossel and Baldur Eliasson (ABB Switzerland Ltd.), "Energy and the Hydrogen Economy," January 2003, www.idatech.com/solutions/multi_fuel_solutions/Hydrogen%20Economy%20Report%202003.pdf.

12. Raymond Drnevich, "Hydrogen Delivery: Liquefaction & Compression" (presentation to the H_2 Delivery and Infrastructure Workshop, Washington, DC, May 7–8, 2003), www.eere.energy.gov/hydrogenandfuelcells/pdfs/liquefaction_comp_pres_praxair.pdf.

13. Bossel and Eliasson, "Energy and the Hydrogen Economy"; Raymond Drnevich, personal communications. Drnevich notes, "If you are looking for the energy needed to compress hydrogen for storage on board a vehicle, it is necessary to consider the overpressure required to fill the vehicle to the desired end pressure in a reasonable period of time." For 5,000 psi onboard storage, this might require on overpressure of 7,000 psi.

14. Thomas and Keller, "Hydrogen Storage—Overview."

15. Dale Simbeck and Elaine Chang (SFA Pacific Inc.), "Hydrogen Supply: Cost Estimate for Hydrogen Pathways—Scoping Analysis" (Golden, CO: U.S. Department of Energy, National Renewable Energy Laboratory, July 2002).

16. Raymond Drnevich, personal communications.

17. James J. Eberhardt, "Fuels of the Future for Cars and Trucks" (presentation to the 2002 Diesel Engine Emissions Reduction [DEER] Conference, San Diego, CA, August 25–29, 2002), www.orau.gov/deer/DEER2002/presentations.htm.

18. Michael Valenti, "Fill 'er Up—with Hydrogen," *Mechanical Engineering* (February 2002), www.memagazine.org/backissues/feb02/features/fillerup/fillerup.html. See also Thomas and Keller, "Hydrogen Storage—Overview," which describes a metal hydride system carrying 3.2 kg of hydrogen and weighing 320 kg.

19. Darlene Slattery and Michael Hampton, "Complex Hydrides for Hydrogen Storage," in *Hydrogen, Fuel Cells, and Infrastructure Technologies: FY 2002 Progress Report*, Sec-

tion III.C.2 (Washington, DC: U.S. Department of Energy, November 2002), www.eere.energy.gov/hydrogenandfuelcells/pdfs/33098_sec3.pdf.

20. R. O. Loutfy and E. M. Wexler, "Feasibility of Fullerene Hydride as a High Capacity Hydrogen Storage Material," *Proceedings of the 2001 DOE Hydrogen Program Review* (Golden, CO: U.S. Department of Energy, National Renewable Energy Laboratory, 2001), www.eere.energy.gov/hydrogenandfuelcells/pdfs/30535al.pdf.

21. California Fuel Cell Partnership (CAFCP), "Bringing Fuel Cell Vehicles to Market: Scenarios and Challenges with Fuel Alternatives," report prepared by Bevilacqua Knight Inc., October 2001, www.fuelcellpartnership.org/documents/ScenarioStudy_v1-1.pdf.

22. National Research Council (NRC), *Novel Approaches to Carbon Management: Separation, Capture, Sequestration, and Conversion to Useful Products—Workshop Report* (Washington, DC: National Academies Press, 2003), www.nap.edu/books/0309089379/html/.

23. "Toyota to Recall Fuel Cell Cars," *Automotive News Europe* (May 20, 2003), www.autonews.com/news.cms?newsId=5370.

24. "Toyota Recalls Fuel-Cell Cars Due to Hydrogen Leak," *Agence France-Presse* (May 20, 2003).

25. Dale Simbeck (SFA Pacific Inc.), "CO_2 Capture and Storage—the Essential Bridge to the Hydrogen Economy" (paper presented at the Sixth International Conference on Greenhouse Gas Control Technologies [GHGT-6], Kyoto, Japan, October 1–4, 2002).

26. Marianne Mintz et al., "Hydrogen: On the Horizon or Just a Mirage?" (presentation to the 2002 Future Car Congress, Arlington, VA, June 2002); and Simbeck, "CO_2 Capture and Storage."

27. Mintz et al., "Hydrogen."

28. Ibid.

29. James Tobin, "Natural Gas Transportation—Infrastructure Issues and Operational Trends" (Washington, DC: U.S. Department of Energy, Energy Information Administration, Natural Gas Division, October 2001), www.eia.doe.gov/pub/oil_gas-natural_gas/analysis_publications/natural_gas_infrastructure_issue/pdf/nginfrais.pdf.

30. Mintz et al., "Hydrogen." See also Dale Simbeck (SFA Pacific Inc.), "Hydrogen Production and Infrastructure Costs" (presentation to the Energy Frontiers International Fuels and Engines Conference, Charleston, SC, February 2003).

31. Drnevich, "Hydrogen Delivery."

32. Bossel and Eliasson, "Energy and the Hydrogen Economy." The canister trucks carry 500 kg of hydrogen but deliver only 400 kg. These canisters typically don't discharge all their fuel because extra compressors would be needed to completely empty the delivery tank (once the pressure in the delivery tank dropped below the pressure in the receiving tank), complicating the process and necessitating more capital equipment at the fueling station.

33. The possibility of delivering high-pressure tanks that could be directly installed in cars seems theoretically attractive but in practice would be quite difficult.

34. The discussion of the Hindenburg crash and Bain's theory is based on A. Bain and W. D. van Vorst, "The Hindenburg Tragedy Revisited: The Fatal Flaw Found," *International Journal of Hydrogen Energy* 24 (1999): 399–403; Addison Bain and Ulrich Schmidtchen, "Afterglow of a Myth" (Berlin: German Hydrogen Association, January

2000), www.dwv-info.de/pm/hindbg/hbe.htm; and U.S. Department of Energy, Office of Energy Efficiency and Renewable Energy, "The Hindenburg Myth," www.eere.energy.gov/hydrogenandfuelcells/codes/safety_feature.html.

35. Mariette DiChristina, "What Really Downed the Hindenburg?" *Popular Science* 251, no. 5 (November 1, 1997). See also Russell Moy, "Tort Law Considerations for the Hydrogen Economy," *Energy Law Journal* (November 2003).

36. Arthur D. Little Inc. (ADL), *Guidance for Transportation Technologies: Fuel Choice for Fuel-Cell Vehicles, Final Report,* Phase II Final Deliverable to DOE, 35340-00 (Cambridge, MA: ADL, February, 2002), appendix, p. 107, www.cartech.doe.gov/pdfs/FC/192.pdf.

37. James Hansel (Air Products and Chemicals Inc.), "Safety Considerations for Handling Hydrogen" (presentation to the Ford Motor Company, Allentown, PA, June 12, 1998).

38. Simbeck, "CO_2 Capture and Storage."

39. ADL, *Guidance for Transportation Technologies.*

40. Jim Campbell (Air Liquide), "Hydrogen Delivery Technologies and Systems: Pipeline Transmission of Hydrogen" (presentation to the Strategic Initiatives for Hydrogen Delivery Workshop, Washington, DC, May 7–8, 2003), www.eere.energy.gov/hydrogenandfuelcells/pdfs/hydrogen_pipelines_pres_air_liquid.pdf.

41. "Hydrogen Material Safety Data Sheet," in Hansel, "Safety Considerations for Handling Hydrogen."

42. Hansel, "Safety Considerations for Handling Hydrogen," p. 27 (emphasis added).

43. National Aeronautics and Space Administration (NASA), Office of Safety and Mission Assurance, *Safety Standard for Hydrogen and Hydrogen Systems* (Washington, DC: NASA, 1997), pp. 6-7, 6-8.

44. The 22 percent figure comes from Moy, "Tort Law," whereas Hansel, "Safety Considerations for Handling Hydrogen," gives a 40 percent figure.

45. Moy, "Tort Law." See also Russell Moy, "Liability and the Hydrogen Economy," *Science* 31 (July 4 2003): 47.

46. Valenti, "Fill 'er Up."

47. Campbell, "Hydrogen Delivery Technologies and Systems."

48. DailyBulletin.Com, "H_2 Tanker Ignites in California; Fire Gradually Extinguishing Itself," May 21, 2003, www.evworld.com/databases/shownews.cfm?pageid=news210503-02. The tanker "was shooting out an orange flame through a pressure release valve." Interestingly, while hydrogen flames are generally invisible, a 2003 report noted that flames from high-pressure hydrogen tanks can "have a discernible orange glow. This colour is believed to be due to naturally occurring particulate matter in the air being entrained into the flame." See L. Shirvill et al. (Shell Global Solutions, Chester, England), "Hydrogen Safety Research" (presentation to the National Hydrogen Association's Fourteenth Annual U.S. Hydrogen Conference and Hydrogen Expo, Washington, DC, March 4–6, 2003).

49. Malcolm A. Weiss et al., "Comparative Assessment of Fuel Cell Cars" (Cambridge, MA: Massachusetts Institute of Technology, 2003), http://lfee.mit.edu/publications/PDF/LFEE_2003-001_RP.pdf.

50. See, for instance, ADL, *Guidance for Transportation Technologies.*

51. The energy and greenhouse gas benefits would be higher for methanol that comes from so-called stranded natural gas in remote locations that can be cost-effectively converted to liquid fuel. At the same time, though, such stranded natural gas could also be used to make other kinds of liquid fuel, including liquefied natural gas (LNG), which can be used more efficiently to generate electricity than to power a hydrogen car (see Chapter 8).

52. Joan Ogden et al., "Fuels for Fuel Cell Vehicles," *Fuel Cells Bulletin*, no. 16 (January 2000).

53. Ibid.

54. Ibid.

55. "Shell Calls for Public Debate to Ensure Hydrogen Becomes the Cheaper, Cleaner Fuel of the Future," Shell Hydrogen press release, September 13, 2000, www.shell.com/home/Framework?siteId = hydrogen-en&FC1 = &FC2 = &FC4 = &FC5 = &FC3 = /hydrogen-en/html/iwgen/news_and_library/pressreleases/2000/publicdebate_10040810.html.

56. Sean Casten, Peter Teagan, and Richard Stobart (Arthur D. Little Inc.), "PEM Fuel Cell Technology: The Best Choice for Fuel Cell Powered Vehicles" (presentation to the IQPC Conference, Cambridge, MA, July 25, 2000), www.cambridgeconsultants.com/PDFs/fuelCells00.pdf.

57. CaFCP, "Bringing Fuel Cell Vehicles to Market," pp. 4-16.

58. Ibid., pp. 4-23.

59. Ibid., pp. 4-24.

Chapter 6: The Long Road to Commercialization of Fuel Cell Vehicles

1. Chester Dawson, "Fuel Cells: Japan's Carmakers Are Flooring It," *Business Week* (December 23, 2002): 50.

2. Arthur D. Little Inc. (ADL), *Guidance for Transportation Technologies: Fuel Choice for Fuel-Cell Vehicles, Final Report*, Phase II Final Deliverable to DOE, 35340-00 (Cambridge, MA: ADL, February, 2002), p. 31, www.cartech.doe.gov/pdfs/FC/192.pdf.

3. Don Huberts, quoted in "Looking Ahead: Fuel Producers Weigh In on Hydrogen's Fit in Cleaner Energy Production," *Fuel Cell Industry Report* (January 2003), www.fcellreport.com/fcir/fcirreport.asp.

4. D. B. Myers et al. (Directed Technologies Inc.), "Economic Comparison of Renewable Sources for Vehicular Hydrogen in 2040" (presentation to the National Hydrogen Association's Fourteenth Annual U.S. Hydrogen Conference and Hydrogen Expo, Washington, DC, March 4–6, 2003).

5. Poul Erik Bak, "Q&A with Mr. S. David Freeman on Hydrogen Now," *H2CarsBiz* (June 15, 2003), www.h2cars.biz/artman/publish/article_181.shtml.

6. Some promote a hydrogen ICE hybrid electric vehicle as the vehicle for market entry. See, for instance, Sandy Thomas, "Hydrogen-Fueled Vehicles: Hybrids vs. Fuel Cells" (presentation to the International Vehicle Technology Symposium on Climate Change, California Air Resources Board, Sacramento, CA, March 2003). Ford is testing such a vehicle, the H²RV. See "H²RV—Ford Hydrogen Hybrid Research Vehicle," Ford press release, August 10, 2003, http://media.ford.com/article_display.cfm?article_id = 16082.

Thomas told me in mid-2003 that he now thinks a hydrogen internal combustion engine hybrid is a better way to go than a hydrogen fuel cell vehicle. Yet Ford's vehicle (based on Ford's Focus wagon) has a fuel economy of 45 miles per kilogram (kg) of hydrogen, whereas the 2004 Toyota Prius (which weighs about 15 percent less than Ford's wagon) gets 20 percent better mileage on an equivalent energy basis, about 55 miles per gallon of gasoline. If a kilogram of hydrogen costs three to four times as much as a gallon of gasoline, as it will for the foreseeable future, the H^2RV will be very expensive to drive. With a 5,000 psi tank, the H^2RV has a driving range of 125 miles, whereas the Prius has a range of more than 500 miles.

7. ADL, *Guidance for Transportation Technologies*.

8. Frank Kreith et al., "Legislative and Technical Perspectives for Advanced Ground Transportation Systems," *Transportation Quarterly* 56, no. 1 (Winter 2002): 51–73.

9. Kevin Nesbitt and Daniel Sperling, "Myths Regarding Alternative Fuel Vehicle Demand by Light-Duty Vehicle Fleets," *Transportation Research Part D: Transport and Environment* 3, no. 4 (1998): 259–269, http://repositories.cdlib.org/cgi/viewcontent.cgi?article=1043&context=itsdavis.

10. Ibid.

11. Ibid.

12. Ibid.

13. U.S. General Accounting Office (GAO), *Energy Policy Act of 1992: Limited Progress in Acquiring Alternative Fuel Vehicles and Reaching Fuel Goals*, GAO/RCED-00-59 (Washington, DC: GAO, February 2000), www.ccities.doe.gov/pdfs/GAO progress.pdf.

14. Ibid.

15. Jeremy Rifkin, *The Hydrogen Economy: The Creation of the Worldwide Energy Web and the Redistribution of Power on Earth* (New York: J. P. Tarcher/Putnam, 2002), p. 208. Amory Lovins was one of the first to popularize this notion. For a recent statement of his views, see Amory Lovins, "Twenty Hydrogen Myths" (Old Snowmass, CO: Rocky Mountain Institute, July 2003), www.rmi.org/images/other/E-20HydrogenMyths.pdf.

16. Willett Kempton et al., "Vehicle-to-Grid Power: Battery, Hybrid, and Fuel Cell Vehicles as Resources for Distributed Electric Power in California," report prepared for California Air Resources Board, California Environmental Protection Agency, and Los Angeles Department of Water and Power, June 2001, www.udel.edu/V2G/V2G-Cal-2001.pdf.

17. Most home users would need to find a way to vent the vast majority of this heat, especially in the summertime, to prevent their garages from overheating.

18. Sandy Thomas has proposed that building owners could use a PEM with an oversized reformer that would generate excess hydrogen for use by a PEM car. Yet, as we have seen, stationary PEMs probably do not make much sense for residential use. And to fuel PEM cars using high-pressure hydrogen tanks, the likely storage means for the foreseeable future, the homeowner or building owner would also need an expensive and energy-intensive compressor. It is difficult to see how this could be a practical strategy for the foreseeable future. See Sandy Thomas, "Hydrogen and Fuel Cells: Pathway to a Sustainable Energy Future" (Alexandria, VA: H2Gen Innovations 2002), pp. 19–22, http://66.160.67.66/PDF_Documents/whitepaper.pdf.

19. Alec Brooks and Tom Gage, "Integration of Electric Drive Vehicles with the Elec-

tric Power Grid—a New Value Stream" (presentation to the Eighteenth International Electric Vehicle Symposium and Exhibition, Berlin, October 2001), www.acpropulsion.com/EVS18/ACP_V2G_EVS18.pdf.

20. The main competition to fuel cell vehicles, hybrid electric vehicles (see Chapter 8), could also be plugged into the grid to provide spinning reserves or peak power, since hybrids have electric batteries. DaimlerChrysler has developed a hybrid vehicle for the military, which, when parked, "converts to an electric generator to provide 12.5 kW of continuous electric power, or up to 30 kW peak electric power." See DaimlerChrysler, "Hybrids: Environmentally Friendly Predecessors to Fuel Cells," *AFV Quarterly* (Fall 2002).

Chapter 7: *Global Warming and Scenarios for a Hydrogen Transition*

1. Executive Office of the President, Office of Science and Technology Policy (OSTP), "Climate Change: State of Knowledge" (Washington, DC: OSTP, October 1997).

2. 10 Downing Street, Prime Minister's Office, "Prime Minister's Speech on Sustainable Development," February 2003, www.number-10.gov.uk/output/Page3073.asp.

3. Ibid.

4. Natural Resources Defense Council, "New Science on Global Warming: A Summary of Recent Studies of the Changing Global Climate," www.nrdc.org/globalWarming/fgwscience.asp; and World Wildlife Fund, "Tackling the Problem of Global Warming," www.worldwildlife.org/climate.

5. World Meteorological Organization, *World Climate News,* no. 23 (June 2003): 4, www.wmo.ch/web/catalogue/New%20HTML/frame/engfil/wcn/wcn23.pdf.

6. Michael E. Mann and Philip D. Jones, "Global Surface Temperatures over the Past Two Millennia," *Geophysical Research Letters* 30, no. 15, 1820 (August 2003). The abstract and key figures from the paper are available at www.ngdc.noaa.gov/paleo/pubs/mann2003b/mann2003b.html.

7. U.S. Department of State, Bureau of Oceans and International Environmental and Scientific Affairs, "Coral Bleaching, Coral Mortality, and Global Climate Change," report to the U.S. Coral Reef Task Force, March 5, 1999, www.state.gov/www/global/global_issues/coral_reefs/990305_coralreef_rpt.html.

8. T. P. Hughes et al., "Climate Change, Human Impacts, and the Resilience of Coral Reefs," *Science* 301 (August 15, 2003): 929–933.

9. See, for instance, Thomas R. Karl et al., "Trends in U.S. Climate during the Twentieth Century," *Consequences* 1, no. 1 (Spring 1995), www.gcrio.org/CONSEQUENCES/spring95/Climate.html.

10. World Meteorological Organization (WMO), "According to the World Meteorological Organization, Extreme Weather Events Might Increase," press release WMO-No 695, July 2, 2003, www.wmo.ch/web/Press/Press695.doc.

11. "Reaping the Whirlwind: Extreme Weather Prompts Unprecedented Global Warming Alert," *Independent* (July 3, 2003), http://news.independent.co.uk/world/environment/story.jsp?story=421166.

12. Cable News Network, "French Heat Toll Tops 11,000," August 29, 2003,

www.cnn.com/2003/WORLD/europe/08/29/france.heatdeaths; WMO, "Extreme Weather Events Might Increase."

13. An additional 4 billion tons of CO_2 is released annually as a result of land-use changes (mainly burning and decomposition of forest biomass). See Alison Bailie, Stephen Bernow, Brian Castelli, Pete O'Connor, and Joseph Romm, *The Path to Carbon Dioxide-Free Power: Switching to Clean Energy in the Utility Sector,* report for the World Wildlife Fund (Washington, DC: Tellus Institute and Center for Energy and Climate Solutions, April 2003), www.worldwildlife.org/powerswitch/power_switch.pdf.

14. National Research Council (NRC), *Climate Change Science: An Analysis of Some Key Questions* (Washington, DC: National Academy Press, 2001), www.nap.edu/html/climatechange.

15. Royal Commission on Environmental Pollution, *Energy—the Changing Climate,* report to Parliament, June 2000, Chap. 2, "Causes and Effects of Climate Change," pp. 22–23, www.rcep.org.uk/pdf/chp2.pdf.

16. OSTP, "Climate Change: State of Knowledge."

17. Intergovernmental Panel on Climate Change, *Climate Change 2001: The Scientific Basis* (Cambridge, England: Cambridge University Press, 2001), www.grida.no/climate/ipcc_tar/.

18. Thomas J. Crowley, "Causes of Climate Change over the Past 1000 Years," *Science* 289 (July 2000): 270–277. Crowley analyzed climate data over the past 1,000 years and concluded that "natural variability [played] only a subsidiary role in the 20th century warming and that the most parsimonious explanation for most of the warming is that it is due to the anthropogenic increase in greenhouse gasses."

19. NRC, *Climate Change Science* (emphasis added).

20. Donald Kennedy, "An Unfortunate U-Turn on Carbon," *Science* 291, no. 5513 (March 30, 2001): 2515, www.eletra.com/temi/e_article000019013.cfm. See also S. Fred Singer, "Editor Bias on Climate Change?" and Donald Kennedy's response, *Science* 301, no. 5633 (August 1, 2003), www.sciencemag.org/cgi/content/full/301/5633/595b?etoc.

21. NRC, *Climate Change Science.*

22. Stephen Schneider, "What Is Dangerous Climate Change?" *Nature* 411 (2001): 17–19.

23. Tom Wigley, letter to Senators Tom Daschle and Bill Frist, July 29, 2003.

24. The debate about climate in the 1970s is sometimes mischaracterized as a scientific consensus that the Earth was cooling. Rather, it was not clear at the time whether dust and aerosols injected into the atmosphere by humans would create a cooling effect that would overwhelm the well-understood warming effect created by human generation of CO_2. See, for instance, Tom Alexander, "Ominous Changes in the World's Weather," *Fortune* (February 1974): 89–95, 142, 146, 150, 152. Three decades' worth of climate data and intense climate research, however, has settled this issue for the vast majority of climate scientists, as discussed in the text.

25. Fred Pearce, "Global Warming's Sooty Smokescreen Revealed," *New Scientist* (June 4, 2003), www.newscientist.com/news/news.jsp?id=ns99993798.

26. See also Chris Jones et al., "Strong Carbon Cycle Feedbacks in a Climate Model with Interactive CO_2 and Sulphate Aerosols," *Geophysical Research Letters* 30 (2003): 1479–1482. Here, scientists from the Hadley Centre for Climate Prediction and Research

in the United Kingdom argue that better modeling of the effects of CO_2 and sulfate aerosols yields a projected average global temperature increase of 5.5°C by 2100.

27. Schneider, "What Is Dangerous Climate Change?"

28. Brian C. O'Neill and Michael Oppenheimer, "Dangerous Climate Impacts and the Kyoto Protocol," *Science* 296, no. 5573 (June 14, 2002): 1971–1972, www.mindfully.org/Air/2002/Dangerous-Climate-Impacts-Kyoto14jun02.htm. The quoted text is from "Preserving Options: Short-Term Action Required to Avoid Long-Term Climate Damage," Princeton University press release on this article, June 13, 2002, www.princeton.edu/pr/news/02/q2/0614-warming.htm.

29. Modeling by the Hadley Centre in 2003 revealed that climate feedbacks, such as "loss of soil carbon due to enhanced soil respiration as the climate warms," could yield a CO_2 concentration of 980 ppmv by 2100. See Jones et al., "Strong Carbon Cycle Feedbacks."

30. National Research Council (NRC), *Abrupt Climate Change: Inevitable Surprises* (Washington, DC: National Academies Press, 2002), www.nap.edu/books/0309074347/html. See also Jonathan Adams, Mark Maslin, and Ellen Thomas, "Sudden Climate Transitions during the Quaternary," *Progress in Physical Geography*, vol. 23 (1999), www.esd.ornl.gov/projects/qen/transit.html.

31. NRC, *Abrupt Climate Change.*

32. Ibid.

33. Ibid.

34. John Browne, "Addressing Global Climate Change" (speech delivered at Stanford University, Stanford, CA, May 19, 1997), www.bp.com/centres/press/s_detail.asp?id=29.

35. Sir Philip Watts (Royal Dutch/Shell Group), "Prudence Pays—Practical Steps to Bridge Conflicting Views on Climate Change" (speech delivered at Rice University, Houston, TX, March 12, 2003), www.ruf.rice.edu/~eesi/sustain/Watts.pdf.

36. *Economist* (September 28, 1991). See also Peter Senge, *The Fifth Discipline* (New York: Doubleday, 1990), p. 181.

37. Unless otherwise indicated, quotations in the sections on Shell's scenarios are from Shell International, Global Business Environment, "Energy Needs, Choices, and Possibilities: Scenarios to 2050," 2001, www.shell.com/static/media-en/downloads/51852.pdf.

38. Other questionable assertions and assumptions include, "Fuel cell cars can also be used with 'docking stations' to provide energy to homes and buildings. The US vehicle fleet, with more installed power capacity than the current global electricity industry, is put to better use," and "After 2025 the growing use of fuel cells as a heat and power source creates a rapidly expanding demand for hydrogen."

39. Jeroen van der Veer (Royal Dutch/Shell Group), "Hydrogen—Fuel of the Future" (remarks at Iceland Hydrogen Economy Conference, April 24, 2003), www.shell.com/static/hydrogen-en/downloads/news_and_library/speeches/speech_JvdVeer_2.pdf.

40. Donald P. H. Huberts, "The Path to a Hydrogen Economy," testimony before the House Committee on Science, March 5, 2003, www.house.gov/science/hearings/full03/mar05/huberts.htm.

41. U.S. Department of Energy, Office of Fossil Energy, "FutureGen: A Sequestration and Hydrogen Research Initiative," fact sheet, February 2003, www.energy.gov/engine/doe/files/import/FutureGenFactSheet.pdf.

42. U.S. Department of Energy, Office of Fossil Energy, "Carbon Sequestration R&D Overview," www.fe.doe.gov/programs/sequestration/overview.shtml. Since coal emits about 1 metric ton of CO_2 per MWh, the cost of sequestering CO_2 from coal electricity is currently about $0.03 to $0.08/kWh.

Some people use carbon rather than CO_2 as a metric. The fraction of carbon in CO_2 is the ratio of their weights. The atomic weight of carbon is 12 atomic mass units, whereas the weight of CO_2 is 44 because it includes two oxygen atoms that each weigh 16. So, to switch from one to the other, use the following formula: One ton of carbon equals $44/12 = 11/3 = 3.67$ tons of CO_2. Thus, 11 tons of CO_2 equals 3 tons of carbon, and a price of $30 per ton of CO_2 equals a price of $110 per ton of carbon.

Chapter 8: Coping with the Global Warming Century

1. See, for instance, Eban Goodstein, "Polluted Data," 1997, *The American Prospect*, Vol. 8, No. 35, November–December, 1997, www.prospect.org/print/V8/35/goodstein-e.html, and Winston Harrington, Richard D. Morgenstern, and Peter Nelson, "On the Accuracy of Regulatory Cost Estimates," *Journal of Policy Analysis and Management*, Vol. 19, No.2 (Spring 2000), pp. 297–322.

All references to "tons" in the text refer to metric tons unless otherwise specified.

2. Interlaboratory Working Group on Energy-Efficient and Low-Carbon Technologies (Oak Ridge National Laboratory, Lawrence Berkeley National Laboratory, Argonne National Laboratory, National Renewable Energy Laboratory, and Pacific Northwest National Laboratory), *Scenarios of U.S. Carbon Reductions: Potential Impacts of Energy-Efficient and Low-Carbon Technologies by 2010 and Beyond*, report prepared for U.S. Department of Energy, Office of Energy Efficiency and Renewable Energy, LBNL-40533 or ORNL/CON-444, September 1997, www.ornl.gov/ORNL/Energy_Eff/lab-web.htm. For a further discussion of this study, see Arthur Rosenfeld, "The Art of Energy Efficiency: Protecting the Environment with Better Technology," *Annual Review of Energy and the Environment* 24 (1999): 33–82, www.energy.ca.gov/commission/commissioners/2000-10_ROSENFELD_AUTOBIO.PDF.

3. Interlaboratory Working Group on Energy-Efficient and Clean Energy Technologies (Oak Ridge National Laboratory, Lawrence Berkeley National Laboratory, Argonne National Laboratory, National Renewable Energy Laboratory, and Pacific Northwest National Laboratory), *Scenarios for a Clean Energy Future*, report prepared for U.S. Department of Energy, Office of Energy Efficiency and Renewable Energy, ORNL/CON-476 and LBNL-44029, November 2000, www.ornl.gov/ORNL/ Energy_Eff/CEF.htm.

4. James McVeigh et al., "Winner, Loser, or Innocent Victim? Has Renewable Energy Performed as Expected?" Discussion Paper 99-28 (Washington, DC: Resources for the Future, June 1999), www.rff.org/rff/Documents/RFF-DP-99-28.pdf; and Dallas Burtraw et al., "Renewable Energy: Winner, Loser, or Innocent Victim?" *Resources*, no. 135 (Spring 1999): 9–13, www.rff.org/rff/Documents/RFF-Resources-135-renewenergy.pdf. The renewables were biomass, geothermal, solar photovoltaics, solar thermal, and wind.

5. Linda R. Cohen and Roger G. Noll, eds., *The Technology Pork Barrel* (Washington, DC: Brookings Institution, 1991), p. 363.

6. R. Brent Alderfer, Thomas J. Starrs, and M. Monika Eldridge, "Making Connections: Case Studies of Interconnection Barriers and Their Impact on Distributed Power Projects" (Golden, CO: U.S. Department of Energy, National Renewable Energy Laboratory, July 2000), www.nrel.gov/docs/fy00osti/28053.pdf.

7. Also, local governments that promote fuel cell vehicles are likely to do so in order to take advantage of their zero tailpipe emissions, which is a major benefit, primarily in polluted cities.

8. Malcolm A. Weiss et al., "Comparative Assessment of Fuel Cell Cars" (Cambridge, MA: Massachusetts Institute of Technology, 2003), http://lfee.mit.edu/publications/PDF/LFEE_2003-001_RP.pdf. See also David Greene and Andreas Schafer, "Reducing Greenhouse Gas Emissions from U.S. Transportation" (Arlington, VA: Pew Center on Global Climate Change, May 2003), www.pewclimate.org.

9. Greene and Schafer, "Reducing Greenhouse Gas Emissions," p. 18.

10. Malcolm Weiss et al., "On the Road in 2020: A Life-Cycle Analysis of New Automobile Technologies" (Cambridge: Massachusetts Institute of Technology, October 2000), http://mit42v.mit.edu/public/In_the_News/el00-003.pdf.

11. Information is from Alliance for Environmental Innovation, "FedEx Express–Alliance for Environmental Innovation Project," www.environmentaldefense.org/alliance/fedex_index.htm.

12. Marianne Mintz et al., "Hydrogen: On the Horizon or Just a Mirage?" (presentation to the 2002 Future Car Congress, Arlington, VA, June 2002).

13. Frank Kreith and colleagues calculate well-to-wheel efficiency for a fuel cell vehicle running on hydrogen from natural gas at only 24.5 percent. Frank Kreith et al., "Legislative and Technical Perspectives for Advanced Ground Transportation Systems," *Transportation Quarterly* 56, no. 1 (Winter 2002): 51–73.

14. Weiss et al., "Comparative Assessment of Fuel-cell Cars."

15. Arthur D. Little Inc. (ADL), *Guidance for Transportation Technologies: Fuel Choice for Fuel-Cell Vehicles, Final Report,* Phase II Final Deliverable to DOE, 35340-00 (Cambridge, MA: ADL, February, 2002), www.cartech.doe.gov/pdfs/FC/192.pdf.

16. Here is a back-of-the-envelope calculation analyzing this point. A 27-mpg car traveling 12,000 miles per year emits 5 metric tons of CO_2 per year, using 25 pounds as the life-cycle CO_2 for a gallon of gasoline. Sharon Thomas and Marcia Zalbowitz, *Fuel Cells—Green Power* (Los Alamos, NM: Los Alamos National Laboratory, 1999), p. 28. A fuel cell vehicle, even with zero-carbon hydrogen fuel, with an annual operating cost (including depreciation) of over $1,000 more (ADL, *Guidance for Transportation Technologies*), reduces CO_2 for more than $200 per metric ton (more than $1,000 for 5 tons). A number of studies suggest that a 50 percent gain in fuel efficiency by advanced ICEs as compared with current ICEs is possible with a net *reduction* in annual operating costs (because of the reduced gasoline bill), thus providing CO_2 savings at no net cost or even a net savings. See Weiss et al., "On the Road in 2020," tables 5.3 and 5.4; Interlaboratory Working Group, *Scenarios of U.S. Carbon Reductions,* pp. 5.44–5.48; and Greene and Schafer, "Reducing Greenhouse Gas Emissions," pp. 13–18.

17. David Keith and Alexander Farrell, "Rethinking Hydrogen Cars," *Science* 301 (July 18, 2003): 315–316.

18. This estimate is derived using Shell's numbers, from Jeroen van der Veer (Royal Dutch/Shell Group), "Hydrogen—Fuel of the Future" (remarks at Iceland Hydrogen Economy Conference, April 24, 2003), www.shell.com/static/hydrogen-en/downloads/news_and_library/speeches/speech_JvdVeer_2.pdf. "We estimate that the initial investment required in the US alone to supply just 2% of cars with hydrogen by 2020 is around $20 billion." That means infrastructure for some 4 million cars and light trucks (Stacy C. Davis and Susan W. Diegel, *Transportation Energy Data Book,* 22nd ed. [Oak Ridge, TN: Oak Ridge National Laboratory, 2002]) will cost $20 billion, yielding $5,000 per car.

19. See Keith and Farrell, "Rethinking Hydrogen Cars"; David Milborrow and Lyn Harrison, "Hydrogen Myths and Renewables Reality," *Windpower Monthly* (May 2003): 47–53; and Nick Eyre, Malcolm Fergusson, and Richard Mills, "Fuelling Road Transport: Implications for Energy Policy," report by the Energy Saving Trust (London), the Institute for European Environmental Policy (London), and the National Society for Clean Air and Environmental Protection (Brighton), November 2002, www.est.org.uk/est/documents/Fueling_Road_Transport_Jan_03.pdf.

20. U.S. Department of Energy, Energy Information Administration (EIA), *Annual Energy Outlook 2003* (Washington, DC: EIA, January 2003), p. 67.

21. Dale E. Heydlauff, personal communications.

22. Alan Greenspan, testimony on natural gas supply and demand issues before the House Committee on Energy and Commerce, June 10, 2003, www.federalreserve.gov/boarddocs/testimony/2003/20030610/default.htm. See also "Greenspan Decries State of Natural-Gas Policy," *Wall Street Journal* (June 10, 2003), http://online.wsj.com/article/0,,SB10552614395278700,00.html.

23. Sustainable Solutions Pty. Ltd., *Strategic Study of Household Energy and Greenhouse Issues,* report prepared for the Australian Greenhouse Office, June 1998, chapter 4, "Greenhouse Intensity of Energy Sources," www.greenhouse.gov.au/coolcommunities/strategic/chapter4.html. See also Michael Wang (Center for Transportation Research, Argonne National Laboratory), "Fuel Choices for Fuel-Cell Vehicles: Well-to-Wheels Energy and Emission Impacts" (presentation to the 2002 Fuel Cell Seminar, Palm Springs, CA, November 2002), www.transportation.anl.gov/pdfs/TA/260.pdf. And see CH-IV International, "LNG Fact Sheet" (Millersville, MD: CH-IVInternational, 2003), www.ch-iv.com/lng/lngfact.htm.

24. This energy penalty does not significantly diminish the tremendous greenhouse gas savings possible from replacing inefficient coal power with efficient natural gas power.

25. David G. Hawkins, testimony before the U.S. House Committee on Energy and Commerce, Subcommittee on Energy and Air Quality, June 24, 2003, www.nrdc.org/globalWarming/tdh0603.asp.

26. Ibid.

27. Eyre, Fergusson, and Mills, "Fuelling Road Transport," pp. 35–38.

28. In the absence of any policies to restrict CO_2 emissions in the electric utility sector, new renewable power would tend to displace mainly natural gas combined cycle plants. However, once a cap on CO_2 is put in place, new renewables (and new natural gas plants) would tend to displace the most carbon-intensive power sources, which are coal-

fired plants. See, for instance, Interlaboratory Working Group, *Scenarios of U.S. Carbon Reductions,* pp. 6.8–6.16.

29. Eyre, Fergusson, and Mills, "Fuelling Road Transport."

30. Keith and Farrell, "Rethinking Hydrogen Cars."

31. U.S. Department of Energy, Energy Information Administration (EIA), *Analysis of Strategies for Reducing Multiple Emissions from Electric Power Plants: Sulfur Dioxide, Nitrogen Oxides, Carbon Dioxide, and Mercury and a Renewable Portfolio Standard,* SR/OIAF/2001-03 (Washington, DC: EIA, July 2001), http://tonto.eia.doe.gov/FTP-ROOT/service/oiaf2001-03.pdf.

32. Milborrow and Harrison, "Hydrogen Myths," p. 52.

33. Ken Caldeira, Atul K. Jain, and Martin I. Hoffert, "Climate Sensitivity Uncertainty and the Need for Energy without CO_2 Emission," *Science* 299, no. 5615 (March 28, 2003): 2052–2054. See also Lawrence Livermore National Laboratory, "Scientists Urge Development of Non-CO_2-Emitting Energy Sources to Avoid Risk of Dangerous Climate Change," news release, March 27, 2003, www.llnl.gov/llnl/06news/NewsReleases/2003/NR-03-03-09.html.

34. Hawkins, testimony before House Committee on Energy and Commerce.

35. Caldeira, Jain, and Hoffert, "Climate Sensitivity Uncertainty."

36. Kreith et al., "Legislative and Technical Perspectives."

37. Keith and Farrell, "Rethinking Hydrogen Cars."

38. President George W. Bush, speech on energy independence at the National Building Museum, Washington, DC, February 6, 2003, excerpted at www.eere.energy.gov/femp/resources/pdfs/021103insights.pdf.

39. Davis and Diegel, *Transportation Energy Data Book,* pp. 5-2, 5-3.

40. National Research Council (NRC), *Effectiveness and Impact of Corporate Average Fuel Economy (CAFE) Standards* (Washington, DC: National Academy Press, 2002), www.nap.edu/books/0309076013/html/.

41. Interlaboratory Working Group, *Scenarios of U.S. Carbon Reductions,* pp. 5.44–5.48; Weiss et al., "On the Road in 2020," tables 5.3 and 5.4; and Greene and Schafer, "Reducing Greenhouse Gas Emissions," pp. 13–18. A study carried out by the Oak Ridge National Laboratory found that "based on a comparison of fatality data for SUVs to other vehicles, the registered-vehicle-fatality rate (defined as number of fatalities per number of registered vehicles) for SUVs is higher than the registered-vehicle-fatality rate for other vehicles." Stacy Davis and Lorena Truett, "An Analysis of the Impact of Sport Utility Vehicles in the United States" (Oak Ridge, TN: Oak Ridge National Laboratory, August 2000), p. 24, www-cta.ornl.gov/cta/Publications/Final%20SUV%20report.pdf.

42. Greene and Schafer, "Reducing Greenhouse Gas Emissions," p. 48.

43. Alliance to Save Energy, "Increasing Automobile Fuel Efficiency," fact sheet (Washington, DC: Alliance to Save Energy, May 2003), www.ase.org/policy/factsheets/TFS.htm. Engine efficiency has increased continuously for three decades; in the late 1970s and early 1980s, manufacturers used part of this improvement to increase fuel economy and part of it to increase engine power; now virtually all the improvements go toward increasing engine power for a fixed fuel economy. See U.S. Environmental Protection Agency (EPA), "Light-Duty Automotive Technology and Fuel Economy Trends:

1975 through 2003," executive summary (Washington, DC: EPA, April 2003), www.epa.gov/otaq/cert/mpg/fetrends/s03004.pdf.

44. Danny Hakim, "Ford Says New S.U.V.'s Less Fuel-Efficient Than Old Ones," *New York Times* (July 18, 2003), www.nytimes.com/2003/07/18/business/18CND-FORD.html.

45. EIA, *Annual Energy Outlook 2003*.

46. Matthew Simmons, address to the Second International Conference of the Association for the Study of Peak Oil, Paris, May 27, 2003, www.fromthewilderness.com/free/ww3/061203_simmons.html.

47. Kenneth Deffeyes, *Hubbert's Peak: The Impending World Oil Shortage* (Princeton, NJ: Princeton University Press, 2001), p. 158.

48. Shell International, Global Business Environment, "Energy Needs, Choices, and Possibilities: Scenarios to 2050," 2001, p. 18, www.shell.com/static/media-en/downloads/51852.pdf.

49. Michael Lynch (responding to a question by Congressman Vernon Ehlers), in "U.S. Energy Outlook and Implications for Energy R&D," hearing before the U.S. House Committee on Science, Subcommittee on Energy and Environment, March 14, 1996, p. 160.

50. U.S. Department of Energy, Office of Fossil Energy, "Carbon Sequestration R&D Overview," www.fe.doe.gov/programs/sequestration/overview.shtml. This is the cost for large-scale sequestration in places such as deep underground aquifers. Small-scale sequestration for enhanced oil and gas recovery is far less expensive.

51. National Research Council (NRC), *Novel Approaches to Carbon Management: Separation, Capture, Sequestration, and Conversion to Useful Products—Workshop Report* (Washington, DC: National Academies Press, 2003), p. 3, www.nap.edu/books/0309089379/html/.

52. "Initially the hydrogen will be used as a clean fuel for electric power generation either in turbines, fuel cells or hybrid combinations of these technologies." U.S. Department of Energy, Office of Fossil Energy, "FutureGen: A Sequestration and Hydrogen Research Initiative," fact sheet, February 2003, www.energy.gov/engine/doe/files/import/FutureGenFactSheet.pdf.

53. D. Gray and G. Tomlinson, "Hydrogen from Coal," Mitretek Technical Paper MTR 2002-31, report for U.S. Department of Energy, National Energy Technology Laboratory (NETL), July 2002, www.netl.doe.gov/coalpower/gasification/pubs/pdf/HYDROGEN%20FROM%20COAL4.pdf.

54. NRC, *Novel Approaches to Carbon Management*.

55. J. Dooley and M. Wise, "Why Injecting CO_2 into Various Geologic Formations Is Not the Same as Climate Change Mitigation: The Issue of Leakage" (College Park, MD: Joint Global Change Research Institute [Battelle–Pacific Northwest National Laboratory], 2002). See also David Hawkins, "Passing Gas: Policy Implications of Leakage from Geologic Carbon Storage Sites" (Washington, DC: Natural Resources Defense Council, 2002).

56. Robert H. Williams, "Decarbonized Fossil Energy Carriers and Their Energy Technology Competitors" (Princeton, NJ: Princeton University, Princeton Environmental Institute, 2003), www.nrcan.gc.ca/es/etb/cetc/combustion/co2network/pdfs/ipcc_implications.pdf.

57. Dale Simbeck, "Gasification Repowering: The Innovative Option for Old Existing Coal-Fired Power Plants" (presentation to the eleventh annual National Energy Modeling System/Annual Energy Outlook Conference, Washington, DC, March 18, 2003). Simbeck proposes retrofitting existing coal plants with gasification and sequestration technology rather than building entirely new plants.

58. U.S. Department of Energy, "FutureGen," p. 2.

59. National Coal Council, *Coal Related Greenhouse Gas Management Issues* (Washington, DC: National Coal Council, May 2003), www.nationalcoalcouncil.org/Documents/fpb.pdf. See also Simbeck, "Gasification Repowering."

60. See, for instance, Keith and Farrell, "Rethinking Hydrogen Cars"; Timothy Johnson and David Keith, "Fossil Electricity and CO_2 Sequestration," *Energy Policy*, Vol. 32 (2004): 367–382, www.andrew.cmu.edu/user/dk3p/papers/49.Johnson.2004.FossilElectricityWithoutCO_2.e.pdf; and Howard Herzog, "The Economics of CO_2 Separation and Capture," *Technology* 7, suppl. 1 (2000): 13–23, http://sequestration.mit.edu/pdf/economics_in_technology.pdf.

61. Michael Obersteiner et al., "Managing Climate Risk," Interim Report IR-01-051 (Laxenburg, Austria: International Institute for Applied Systems Analysis, December 2001), www.iiasa.ac.at/Publications/Documents/IR-01-051.pdf.

62. See Lee Lynd et al., "Bioenergy: Background, Potential, and Policy," policy briefing prepared for the Center for Strategic and International Studies (CSIS), Washington, DC, January 2003, www.csis.org/tech/biotech/other/Lynd.pdf.

63. Obersteiner et al., in "Managing Climate Risk" (p. 2), note, "Stabilization at a target level is a non-robust strategy in an environment that is extremely uncertain and most likely nonlinear."

64. Lester B. Lave, W. Michael Griffin, and Heather MacLean, "The Ethanol Answer to Carbon Emissions," *Issues in Science and Technology* (Winter 2001), www.nap.edu/issues/18.2/lave.html. See also Lester Lave et al., "Life-Cycle Analysis of Alternative Automobile Fuel/Propulsion Technologies," *Environmental Science and Technology* 34 (2000): 3598–3605.

65. Michael Bryan, "The Fuels Market—Biofuel Penetration and Barriers to Expansion" (presentation to conference, National Security and Our Dependence on Foreign Oil, Center for Strategic and International Studies, Washington, DC, June 2002), pp. 13–15, www.csis.org/tech/biotech/other/Ebel.pdf.

66. Greene and Schafer, "Reducing Greenhouse Gas Emissions," p. 30.

67. Lave, Griffin, and MacLean, "Ethanol Answer."

68. Lynd, personal communications.

69. Shell, "Scenarios," p. 22.

Chapter 9: Hydrogen Partnerships and Pilots

1. Jón Björn Skúlason and Helgi Bjarnason, "Hydrogen Developments: Case of Iceland" (presentation to the International Energy Agency and Organization for Economic Cooperation and Development [IEA/OECD] conference, Towards Hydrogen, Paris, March 2003), www.iea.org/workshop/2003/hydrogen/keynotes/iceland.pdf. Emissions of CO_2 alone are between 2 million and 2.5 million tons per year, with other greenhouse

gas compounds adding about 0.5 million ton of CO_2 equivalent per year. See table 10 of "Iceland's Third National Communication under the United Nations Framework Convention on Climate Change" (Reykjavík, Iceland: Ministry for the Environment, April 2003), http://unfccc.int/text/resource/docs/natc/icenc3.pdf.

2. Max Pemberton, *The Iron Pirate* (London: Greycoine Book Manufacturing Company, n.d.), p. 155.

3. Bragi Árnason and Thorsteinn Sigfusson, "Iceland—a Future Hydrogen Economy," *International Journal of Hydrogen Energy* 25 (2000): 389–394. Iceland does have peat resources, which have been used as fuel in the past. Current consumption is reported as zero, according to the World Energy Council's Energy Info Centre, www.worldenergy.org/wec-geis/edc/countries/Iceland.asp.

4. Bragi Árnason et al., "Creating a Non-fossil Energy Economy in Iceland" (Reykjavík, Iceland: Icelandic New Energy Ltd., March 2001).

5. Thorsteinn Sigfusson and Bragi Árnason, "New Perspectives for Renewable Energy in Iceland" (Reykjavík: University of Iceland, Science Institute), 2000, www.newenergy.is/subjects/37/original/video_cover/Subj37-0001.pdf.

6. Árni Ragnarsson, Iceland National Energy Authority (Orkustofnun), personal communications. In some cases in which geothermal energy has been developed for district heating, engineers have added back-pressure turbines to generate electricity.

7. Thorkell Helgason, Egill B. Hreinsson, and Fridrik Sophusson, "Planning and Utilization of Geothermal and Hydroelectric Resources in Iceland" (presentation to ENERGEX 2000, Las Vegas, Nevada, July 23–28, 2000), http://verk.hi.is/~egill/rit/paper%20energex2000.pdf.

8. V. Stefansson, "The Renewability of Geothermal Energy" (presentation to the World Geothermal Congress, Kyushu-Tohoku, Japan, May 28–June 10, 2000), www.geothermie.de/egec-geothernet/prof/0776.PDF.

9. The Wairakei field in New Zealand and the Geysers field in California have both shown declines in steam pressure, which have since been somewhat mitigated by re-injection of water into the reservoir. Allan Clotworthy of Contact Energy Ltd. estimates that Wairakei will be able to produce power at current levels for at least another forty to fifty years (overall lifetime eighty to ninety years). Allan Clotworthy, "Response of Wairakei Geothermal Reservoir to 40 Years of Production" (presentation to the World Geothermal Congress, Kyushu-Tohoku, Japan, May 28–June 10, 2000), www.geothermie.de/egec-geothernet/ci_prof/australia_ozean/new_zealand/0080.PDF.

10. Stefansson, "Renewability of Geothermal Energy." See also the Web site of Iceland's Ministries of Industry and Commerce, http://idnadarraduneyti.is/interpro/ivr/ivreng.nsf/pages/index.html.

11. Stefansson, "Renewability of Geothermal Energy." Assuming an efficiency of 10 percent for power generation, as do Sigfusson and Árnason in "New Perspectives," the 1 million terawatt-hours (TWh) of heat energy could produce about 100,000 TWh of electricity. Waste heat could supply district heating needs.

12. Helgason, Hreinsson, and Sophusson, in "Planning and Utilization," cite a figure of 1,014 MW. A figure of 1,064 MW is cited for 2001 in U.S. Department of Energy, Energy Information Administration, "Country Energy Data Report: Iceland," www.eia.doe.gov/emeu/world/country/cntry_IC.html.

13. Skúlason and Bjarnason, "Hydrogen Developments."

14. See the Kárahnjúkar Hydropower Project's Web site, www.karahnjukar.is/En/category.asp?catID=169. The capacity of the plant is reported in other sources as 500+ MW, 690 MW, or 750 MW.

15. World Wildlife Fund (WWF) and Iceland Nature Conservation Association, "Transition to a Hydrogen Economy—a Strategy for Sustainable Development in Iceland," April 2001, www.panda.org/downloads/europe/hydrogeniniceland.pdf. This report provides an excellent overview of the Icelandic transition, prospects, and potential pitfalls.

16. Ibid.

17. Skúlason and Bjarnason, "Hydrogen Developments."

18. Sigfusson and Árnason, in "New Perspectives," use a figure of 30 TWh per year, equal to 4 GW with an 85 percent capacity factor. Iceland's Ministries of Industry and Commerce use a figure of 25–30 TWh/yr. See also World Energy Council, Energy Info Centre, www.worldenergy.org/wec-geis/edc/countries/Iceland.asp.

19. WWF and Iceland Nature Conservation Association, "Transition to a Hydrogen Economy."

20. "Welfare for the Future: Iceland's National Strategy for Sustainable Development 2002–2020" (Reykjavík, Iceland: Ministry for the Environment, August 2002), http://government.is/interpro/umh/umh-english.nsf/pages/welfare.

21. "Iceland's Third National Communication."

22. Icelandic New Energy Ltd., *ECTOS: Ecological City Transport System*, first newsletter, www.hydrogen.is/subjects/63/original/video_cover/Subj63-0001.pdf.

23. U.S. Department of Energy, Energy Information Administration (EIA), "OECD Total Net Oil Imports," spreadsheet, *International Petroleum Monthly*, www.eia.doe.gov/emeu/ipsr/t47.xls. This shows 2002 imports in million barrels per day.

24. The *Monetary Bulletin: Economic and Monetary Developments and Prospects*, published by the Central Bank of Iceland (www.sedlabanki.is/uploads/files/mb002_3.pdf), shows gasoline prices for 1997–2000. An entry for Iceland in the Central Intelligence Agency's *World Factbook 2003* shows exchange rates for those years (www.cia.gov/cia/publications/factbook/geos/ic.html).

25. Secretariat of the United Nations Framework Convention on Climate Change, "Summary of the Report on the In-Depth Review of the National Communication of Iceland," FCCC/IDR.1(SUM)/ICE, April 7, 1997, http://unfccc.int/cop5/resource/docs/sum/ice01.htm.

26. Ruggero Bertani and Ian Thain, "Geothermal Power Generating Plant CO_2 Survey," *International Geothermal Association News* 49 (July–September 2002), www.geothermie.de/iganews/no49/geothermal_power_generating_plant.htm. Binary-cycle geothermal plants make electricity from moderate-temperature resources, producing negligible emissions and without depleting reservoir fluid, since they do not release reservoir gases or liquids into the environment. However, in Iceland they are used only to a limited extent, since "flash-cycle" plants (which do release emissions) are more efficient at converting energy from Iceland's high-temperature resources.

27. Árnason et al., "Creating a Non-fossil Energy Economy in Iceland."

28. Hydrogen bus demonstration programs will be carried out in nine other Euro-

pean cities as well, providing an opportunity to share lessons and solutions. See ECTOS' Web site, www.hydrogen.is/ectos.asp.

29. Jón Björn Skúlason of Icelandic New Energy Ltd., cited in WWF and Iceland Nature Conservation Association, "Transition to a Hydrogen Economy."

30. Árnason et al., "Creating a Non-fossil Energy Economy in Iceland."

31. The California discussion is based on Alan Lloyd, "The Path to a Hydrogen Economy," testimony submitted to the Committee on Science of the U.S. House of Representatives, Washington, DC, March 5, 2003, www.house.gov/science/hearings/full03/mar05/lloyd.htm; Poul Erik Bak, "California to Establish 'Hydrogen Freeway,'" H2CarsBiz (July 29, 2003), www.h2cars.biz/; and information available at the California Fuel Cell Partnership Web site, www.cafcp.org.

32. The goals are repeated verbatim from California Fuel Cell Partnership (caFCP), "Frequently Asked Questions," April 2003, www.fuelcellpartnership.org/faq.html.

33. Lloyd, "Path to a Hydrogen Economy."

34. California Fuel Cell Partnership (caFCP), "Bringing Fuel Cell Vehicles to Market: Scenarios and Challenges with Fuel Alternatives," report prepared by Bevilacqua Knight Inc., October 2001, www.fuelcellpartnership.org/documents/Scenario Study_vi-1.pdf.

35. Peter Flynn, "Commercializing An Alternate Vehicle Fuel: Lessons Learned from Natural Gas for Vehicles," *Energy Policy* 30, no. 7 (2002): 613–619.

Conclusion: Choosing Our Future

1. National Research Council (NRC), *Abrupt Climate Change: Inevitable Surprises* (Washington, DC: National Academies Press, 2002), www.nap.edu/books/0309074347/html.

2. David Whitford, "Dr. Ballard Thinks Big," *Fortune Small Business* (June 4, 2003), www.fortune.com/fortune/smallbusiness/articles/0,15114,456366,00.html. For a detailed discussion of some of the myriad research problems that must be solved before hydrogen cars will be practical, see U.S. Department of Energy, Office of Science, *Basic Research Needs for the Hydrogen Economy: Report of the Basic Energy Sciences Workshop on Hydrogen Production, Storage, and Use, May 13–15, 2003* (Washington, DC, 2003), www.sc.doe.gov/bes/hydrogen.pdf.

3. See, for instance, "Keystone Dialogue on Global Climate Change," final report (Keystone, CO: Keystone Center, May 2003), www.keystone.org/Public_Policy/FINALREPORTGLOBALCLIMATE.pdf. The report notes: "This analysis led to several key insights. The first is that there is a value associated with early reductions, or conversely, there is a cost associated with delaying reductions."

4. Other designs for an RPS are possible, including ones that include carbon capture and storage or certain kinds of cogeneration.

5. U.S. Department of Energy, Energy Information Administration (EIA), *Analysis of Strategies for Reducing Multiple Emissions from Electric Power Plants: Sulfur Dioxide, Nitrogen Oxides, Carbon Dioxide, and Mercury and a Renewable Portfolio Standard*, SR/OIAF/2001-03 (Washington, DC: EIA, July 2001), http://tonto.eia.doe.gov/FTP-ROOT/service/oiaf2001-03.pdf. Electricity prices in 2020 under a 20 percent RPS would

be about 4 percent higher than the EIA projects they would be in a business-as-usual scenario but 2 percent lower than they are today.

6. Guy Gugliotta and Eric Pianin, "EPA Withholds Air Pollution Analysis," *Washington Post* (July 1, 2003): A03, www.washingtonpost.com/wp-dyn/articles/A54598-2003Jun30.html. See also "Carper-Chafee-Gregg Offer '4 Pollutant Bill': Bipartisan Senators Introduce Clean Air Legislation," press release from Senator Tom Carper, April 28, 2003, http://carper.senate.gov/press/03/04/042803.html.

7. R. Brent Alderfer, Thomas J. Starrs, and M. Monika Eldridge, "Making Connections: Case Studies of Interconnection Barriers and Their Impact on Distributed Power Projects" (Golden, CO: U.S. Department of Energy, National Renewable Energy Laboratory, July 2000), www.nrel.gov/docs/fy00osti/28053.pdf.

8. See, for instance, "Sustainable Development—Reducing Energy Intensity by 2% Per Year" (presentation by Arthur H. Rosenfeld, commissioner, California Energy Commission, to the International Seminar on Planetary Emergencies, Erice, Italy, August 19, 2003), www.energy.ca.gov/commission/commissioners/rosenfeld.html#papers.

9. Joseph J. Romm, *Cool Companies: How the Best Businesses Boost Profits and Productivity by Cutting Greenhouse Gas Emissions* (Washington, DC: Island Press, 1999).

10. Energy Future Coalition, "Challenge and Opportunity: Charting a New Energy Future" (Washington, DC: Energy Future Coalition, June 2003), www.energyfuture-coalition.org/full_report/index.shtm. This estimate incorporates the total energy savings that would result from state and utility energy efficiency programs implemented during 2004–2015, not just the incremental savings from this policy.

11. Alliance to Save Energy, "Why Is Steam Important?" September 2003, www.ase.org/programs/industrial/steam.htm. See U.S. Department of Energy, Office of Industrial Technologies, "Steam," www.oit.doe.gov/bestpractices/steam.

12. See, for instance, Elizabeth Kolbert, "The Car of Tomorrow," *The New Yorker,* August 11, 2003, pp. 36–40.

13. Royal Commission on Environmental Pollution, *Energy—the Changing Climate,* report to Parliament, June 2000, summary, p. 17, www.rcep.org.uk/pdf/enersumm.pdf.

Index